EXPLORING CALCULUS

with

THE GEOMETER'S SKETCHPAD®

CINDY CLEMENTS

RALPH PANTOZZI

SCOTT STEKETEE

Key Curriculum Press
Innovators in Mathematics Education

Project Administrator	Heather Dever
Editorial Assistant	Erin Gray
Reviewers	Christian Aviles-Scott, Steven Chanan, Annie Fetter, Dan Lufkin, Daniel Scher, Anna Werner
Production Editor	Jacqueline Gamble
Copy Editor	Erin Milnes
Production Director	Diana Jean Parks
Editorial Production Manager	Deborah Cogan
Art and Design Coordinator	Caroline Ayres
Cover Designer	Arianna Grabec-Dingman
Cover Photo Credit	Anthony Mercieca/Photo Researchers Inc.
Prepress and Printer	Data Reproductions
Executive Editor	Casey FitzSimons
Publisher	Steven Rasmussen

Exploring Calculus Sketches CD-ROM

Key Curriculum Press guarantees that the *Exploring Calculus* Sketches CD-ROM that accompanies this book is free of defects in materials and workmanship. A defective CD-ROM will be replaced free of charge if returned within 90 days of the purchase date. After 90 days, there is a $10.00 replacement fee.

Key Curriculum Press
1150 65th Street
Emeryville, California 94608
510-595-7000
editorial@keypress.com
http://www.keypress.com

10 9 8 7 6 5 4 3 2 06 05 04 03 02 ISBN 1-55953-535-0

Contents

Introduction

Here and elsewhere we shall not obtain the best insight into things until we actually see them growing from the beginning. Aristotle (*Politics*)

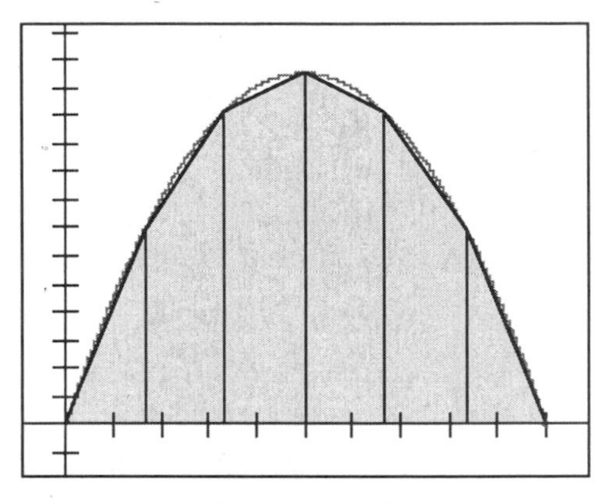

Calculus began as a science of approximations. Archimedes employed the *method of exhaustion* to compute the area enclosed by a parabola. He approximated the area with trapezoids, increasing the number of trapezoids until the area of the parabola was used up. Newton and Leibniz, the "fathers of the calculus," used similar methods to take tentative steps into the unknown, and to speculate about the results.

The Geometer's Sketchpad® makes it possible to view calculus through the same windows as Leibniz and Newton. In this book, you will bring their thought experiments to visual and kinematic life through construction and motion, and rediscover how calculus is built upon a foundation of geometric questions.

Using the Activities

All the activities have been designed with the beginning user of Sketchpad™ in mind. Sketchpad Version 4 has a host of new capabilities and enhancements designed to make the program easier to use, and the sketches for these activities were created to capitalize on these improvements.

Many of the activities can stand alone. The order of the chapters mirrors many calculus texts, but you can also use the activities to explore limits, derivatives, antiderivatives, and integrals at any time in the course.

In Sketchpad, mathematical objects such as function plots, secant lines, and Riemann sums, become animate objects. The activities and their accompanying sketches help students explore the concepts of calculus verbally, numerically, symbolically, graphically—and kinesthetically. The kinesthetic element is a fundamentally new way to explore mathematical ideas, beyond what is possible with concrete manipulatives.

In general, students construct objects where it is mathematically instructive to do so, gaining valuable insights into the structure of their model. They use pre-constructed objects and tools where the construction is less related to the mathematical focus.

The ideas in this book were inspired by the diverse set of learning materials and technologies created over the last decade of change in calculus courses and in mathematics education in general. The activities here are meant to support students' concept building, empowering them to verbalize and symbolize on their own. The activities grew from listening to students as they worked to make calculus "make sense." Students with different strengths—those who are visually or kinesthetically oriented, those who need to see the whole picture before they can start building the parts, and those who prefer to proceed methodically—all have a chance to engage in the problems. Most importantly, you'll honor students' interests and contributions, and help them establish personal connections to mathematics, which in turn strengthens their motivation.

The activities include step-by-step instructions for using tools or constructing objects, and sidebar tips to provide extra assistance. Each activity can be used as a reproducible set of instructions and questions for students to use individually in a computer lab, or with a projection device or set of computers as inspiration for a lesson, presentation, or investigation conducted in a whole-class setting.

Activity Notes

Activity Notes appear at the end of the book. These include suggested prerequisites, answers, explanations of how sketches were constructed, and extensions for those who may have more experience with Sketchpad. They also serve as a home for all the good stuff that just couldn't fit into the main text. You'll find mathematical connections, interesting facts, and pedagogical insights into the structure of the activities.

The Tool Tips document on the CD-ROM provides instructions for creating your own tools to accompany your lessons and classroom investigations. If you're already an experienced Sketchpad user, you can go right to this tools section and start to experiment with the sketches listed. However, we strongly recommend first exploring the activities that are associated with the tools, because you will likely be looking at calculus through "new eyes" as you explore the subject with Sketchpad.

The CD-ROM accompanying the book includes sketches to accompany each activity, as well as sketches that either extend the activities or touch on topics beyond those covered here. You will also find interactive max/min optimization constructions, and demonstrations of how the program can be used with topics ranging from polar coordinates to convergence of series and sequences.

Getting the Most from Written-Response Questions

Although a few of the questions in these activities require short answers from the user (a number or an equation, perhaps), many questions ask for more involved written responses. These questions attempt to guide students into expressing their ideas and hypotheses, reasoning about why certain things happen, and explaining their observations and conclusions. Some students might resist such questions, particularly if they are accustomed to traditional textbook exercises. You can encourage thoughtful writing by providing students with a good reason for it.

Mathematicians communicate through writing to provide clear and enlightening explanations or arguments, to convince each other of the truth and coherence of an argument, and to help themselves think through a problem. Students might be more inclined to write thoughtfully if they are trying to come up with an explanation that can help enlighten a fellow student, or one that most clearly or succinctly explains something. A fruitful discussion might even ensue if students do not all agree on which explanation is the "best." Students can also benefit from working out questions in small groups, or even in whole-class discussions, since they then have to contend with other arguments or counter-arguments.

You can help students who claim not to know where to start on a written response by asking them to describe what they do know, what they can observe, and what they don't understand. This might initiate a student's reasoning process that will at least increase his or her own mathematical awareness, and possibly help answer the question.

Many students are uncomfortable with written responses because they do not know what is expected of them. One way to encourage thoughtful writing is to provide students with a model. Provide a response with some flaws in it, and ask students to critique the argument or to provide a better response. This will give students a better idea of what a good response looks like and will give them practice analyzing good responses.

Acknowledgments

The inspiration and foundation for this book came from Scott Steketee and Nick Jackiw's NCTM workshops entitled "Visualizing Change: Calculus Activities with The Geometer's Sketchpad," and the questions and the ideas of Ralph Pantozzi's students. The writers hope that this book honors Scott and Nick's original vision.

Scott is very grateful to Ralph and Cindy for their contributions, their original ideas, and their hard work to bring this book successfully to publication.

Ralph would like to thank those students, especially Angela, Ankur, Bobby, Brian, Jeff, Magda, Michelle, Mike, Romina, and Sherly for openness and sharing of questions and insights; his wife, Shawna, for understanding of another sort; Bob Davis, who started him thinking about computers and mathematics; Carolyn Maher, who taught him why listening is so important; and his fellow writer, Cindy, for going far beyond the calls of duty.

Cindy would like to thank all the reviewers, as well as Heather Dever, Jackie Gamble, and Karen Howl for their patience, support, and positive feedback, and for keeping her on track, and her cat Bouli for his unconditional love despite her friendliness.

Common Commands and Shortcuts

To open a new sketch
Choose **File | New Sketch.**

To close a sketch
Choose **File | Close,** or click in the close box in the upper-left (Mac) or upper-right (Windows) corner of the sketch.

To undo or redo
Choose **Edit | Undo.** You can undo back to the state your sketch was in when last opened. To redo, choose **Edit | Redo.**

To deselect everything
Click in any blank area of your sketch with the **Arrow** tool or press Esc until objects deselect. To deselect a single object while keeping all other objects selected, click on it with the **Arrow** tool.

To show or hide a label
Position the finger of the **Text** tool over the *object* and click.

To change a label
Position the finger of the **Text** tool over the *label* and double-click.

To change an object's line width or color
Select the object and choose from the appropriate submenu in the Display menu.

To hide an object
Select the object and choose **Display | Hide.**

To plot a function
Choose **Graph | Plot New Function.** Then enter in the expression for the function.

To edit a function (or parameter)
Double-click on the equation for the function (or parameter).

To measure the *x*- or *y*-coordinate of a point
Select the point and choose **Measure | Abscissa (x)** or **Measure | Ordinate (y).**

To calculate the *y*-coordinate of a function
Choose **Measure | Calculate.** Click on the equation for the function, then on the *x*-coordinate for the point.

To plot a point (x_A, y_A) or $(x_A, f(x_A))$
Select measurement x_A, then y_A or $f(x_A)$, and choose **Graph | Plot As (x, y).**

To trace an object
Select the object and choose **Display | Trace.** Do the same thing to toggle off tracing.

To use the Calculator
Choose **Measure | Calculate.** To enter a measurement, click on the measurement itself in your sketch.

To change a parameter to a calculation
Select the parameter and choose **Edit | Edit Parameter.** Enter in the new expression. (*Note:* A parameter is a numerical constant whereas a calculation can contain variables.)

To use a Custom tool
Click on the icon and choose the desired tool from the menu that appears.

Keyboard Shortcuts

Command	Mac	Windows
Undo	⌘+Z	Ctrl+Z
Redo	⌘+R	Ctrl+R
Select All	⌘+A	Ctrl+A
Properties	⌘+?	Alt+?
Hide Objects	⌘+H	Ctrl+H
Show/Hide Labels	⌘+K	Ctrl+K
Trace Objects	⌘+T	Ctrl+T
Erase Traces	⌘+B	Ctrl+B

Command	Mac	Windows
Animate/Pause	⌘+`	Alt+`
Increase Speed	⌘+]	Alt+]
Decrease Speed	⌘+[	Alt+[
Midpoint	⌘+M	Ctrl+M
Intersection	⌘+I	Ctrl+I
Segment	⌘+L	Ctrl+L
Polygon Interior	⌘+P	Ctrl+P
Calculate	⌘+=	Alt+=

Action	Mac	Windows
Scroll drag	Option+ drag	Alt+drag
Display Context menu	Control+ click	Right-click
Navigate Toolbox	Shift+arrow keys	
Choose Arrow, Deselect objects, Stop animations, Erase traces	Esc (escape key)	
Move selected objects 1 pixel	↑, ↓, ←, → keys (Hold down to move continuously)	

1

Exploring Change

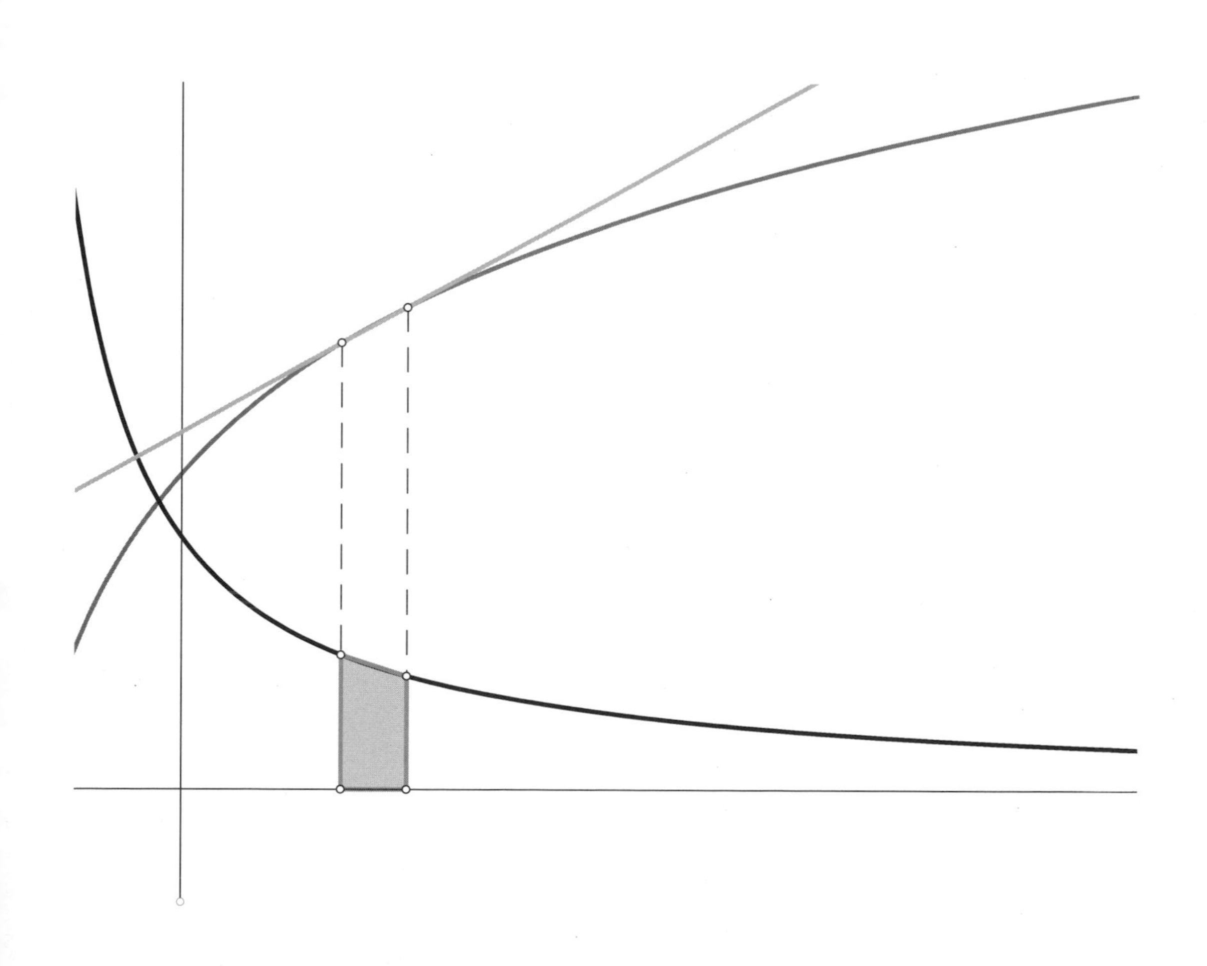

Visualizing Change: Position Name(s): ____________________

Calculus is the study of change and how to model that change—change in amounts, change in position, change in speeds . . . . A familiar situation to start with is distance from a starting point or home base. Suppose you woke up one morning and wanted to describe your travels for the day in terms of distance from your home. How would you do that? Depending on your route, that could get complicated fast. So let's start out easy and assume that every place you want to go, including your home, is conveniently located on a straight line. Now how would you do this?

Sketch and Investigate

One way to begin a good model is to sketch your movement along a line.

1. **Open** the sketch **Position.gsp** in the **Exploring Change** folder. You will see a horizontal line with variable point *Me* which can move between home, school, and other important locations.

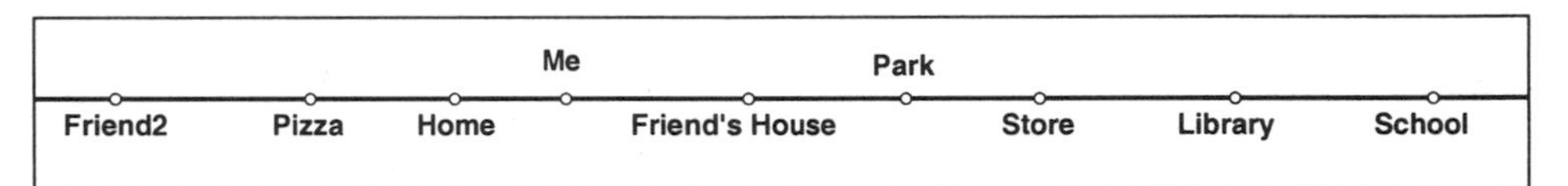

To talk about your *position* along a line in mathematics is to give the distance and direction from the starting point to point *Me*. To calculate your distance along a horizontal line, you subtract the *x*-coordinates of your points.

To measure the *x*-coordinate of a point, select the point with the **Arrow** tool and choose **Abscissa(x)** from the Measure menu.

2. Measure the *x*-coordinate of point *Me*. To calculate the difference $x_{Me} - x_{Home}$ choose **Calculate** from the Measure menu, and then click on each measurement to enter it into the calculator. Label this distance *Position* by choosing the **Text** tool and double-clicking on your measurement, $x_{Me} - x_{Home}$.

To turn on tracing, select point *Me*, then go to the Display menu and choose **Trace Point.**

3. Turn on tracing for point *Me* and drag along your line.

Q1 Can you see your path?

Q2 What happens to your *Position* measurement when you are at a location left of point *Home*? Right of point *Home*? When you are at point *Home*?

The measurement $x_{Me} - x_{Home}$ is called a *directed distance* because it gives both the direction and distance between the two points.

Your trace gives a reasonable model of your path, but it is limited to one dimension. If you look at this trace at the end of the day, can you tell what you did? If you traveled back and forth at all, some information was lost. A better model is a *position plot*, which uses time to expand this line into two dimensions by plotting position versus time. Time is simulated here by the *x*-coordinate of a point moving on an axis.

4. Press the *Show Time* button to show the time axis. Select point *Time* and measure its x-coordinate. Label this measurement *Time.*

5. Select the measurements *Time* and *Position,* in that order, and choose **Plot as (x, y)** from the Graph menu. Label this point P. So the x-coordinate of your point P will be time, and the y-coordinate will be your position.

To change the color or width of an object, select the object, then go to **Color** or **Line Width** in the Display menu.

6. Use the **Arrow** tool to select point P. Color it with a bright color and turn on tracing for point P.

Press the *Wake Up* button to put points *Me, Time,* and P at their starting points. For this first trial, you won't move point *Me* at all.

Q3 What kind of position trace do you think you'll get if point *Me* just sits there and point *Time* moves? (Think! How is your distance from home changing if you stay in your bed as the minutes tick by?)

You can control the speed of animation. Use the Display menu to show the Motion Controller, and then change the speed.

7. Check your answer by pressing the *Animate* button. Observe for awhile, and then press it again to stop.

8. After examining your trace (was it what you expected?) press the *Wake Up* button and choose **Erase Traces** from the Display menu.

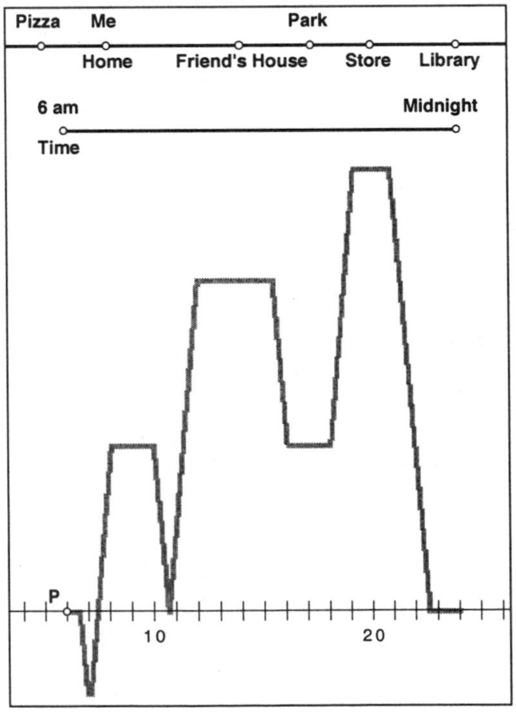

Now drag point *Me* while point *Time* moves. Point P will plot your position with respect to time. You can drag point *Me* any way you wish, but try each of these suggested movements as well. Each time, draw a little sketch of your trace in the margin and make a note of what you did to get that trace. Remember to press *Wake Up* and choose **Erase Traces** when you want to try a new motion.

A. Drag point *Me* at a steady pace to your friend's house (try not to vary your speed at all), stay for about an hour, and then go home at a steady pace.

B. Stay home until 8 AM, and then start out at a slow, steady pace to the right. Speed up gradually until you get to school. Turn around and head toward home at the same quick pace. As you get closer to home, gradually slow down to a crawl.

C. Stay home until 7 AM, and then start out quickly toward school, slowing down more and more as you get closer. Once you get to school, tell the nurse you're sick and start home slowly, speeding up as you get closer to home.

D. Leave home at 7 AM and get to your friend's house about 7:30. Leave for school from your friend's house in time to stop briefly at the store for a snack and still get to school by 8:15. Leave school at 3 PM (15:00), hang out at the store for a half hour, then go to your friend's house for dinner, then off to the library to study for an hour. Make sure you're home by 9 PM (21:00)—it's a school night!

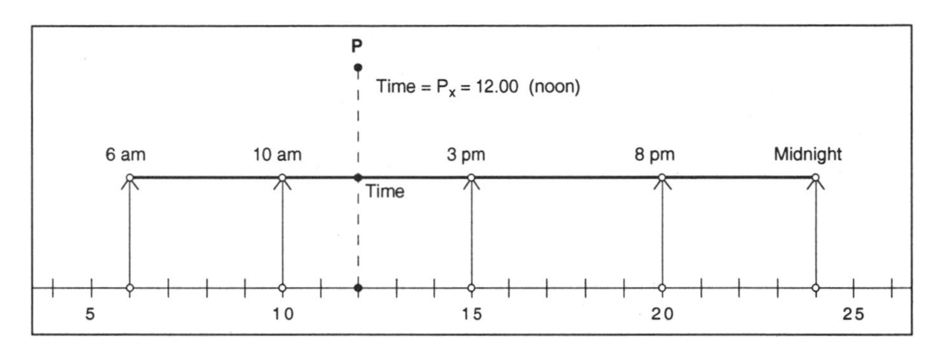

E. Leave home at 9 AM (it's a holiday), pick up Friend1 at your friend's house and go to the store to get picnic supplies. Pick up Friend2 by 11 AM, go get a pizza and head to the park for lunch. Hang out until 2 PM (14:00), then head to the library where they're showing free movies at 3 PM (15:00). Head back home for dinner at 6 PM (18:00).

Make a variety of traces, and look for relationships between how you move the point and the resulting trace.

Q4 How must you drag point *Me* to make the directed distance (position) measurement increase? What happens on the trace when you do this? (Remember that going from –2 to –1 is an increase.)

Q5 How must you drag point *Me* to make the directed distance (position) measurement decrease? What happens on the trace when you do this?

Press the *Saturday* button to show the position graph for a weekend day. Make a trace, trying to match the day's activities as closely as you can. During which part of your travels did you have to go the fastest? When did you move most slowly? When did you stay in one place?

Hide Saturday and press the *Sunday* button. Make a trace, trying to match the day's activities as closely as you can. During which part of your travels did you have to go the fastest? When did you move most slowly? When did you stay in one place? Where did you spend the night?

Explore More

To go to a different page in a sketch, click on the page tab at the bottom left of the window.

Go to page 2 of the sketch. Here you will use the velocity control, rather than dragging point *Me* directly, to control the motion. To move point *Me* and point *P*, press the *Go* button and the *Animate* button. To stop point *Me*, press the *STOP!* button. To stop time, press the *Animate* button again. Experiment with the motion and compare the results of controlling velocity instead of controlling position directly. Try to trace both the Saturday path and the Sunday path. Which is easier to do using the *velocity* slider?

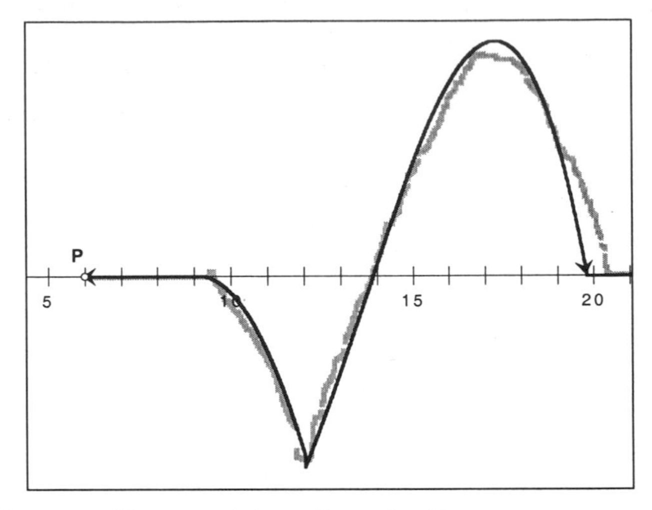

Visualizing Change: Velocity   Name(s): ____________

There are many ways to create motion or move an object. You could control where the object is located—its position—by dragging it around, or you could control how fast or slow the object moves—its speed.

Velocity is related to speed but it provides more information. If you know your velocity, you really know two things—how fast you are moving (speed) and the direction you are heading. Can knowing the velocity of an object tell you anything else? Are there any relationships or patterns between position and velocity? In this activity you will start to answer these questions by moving a point, controlling its velocity with a slider.

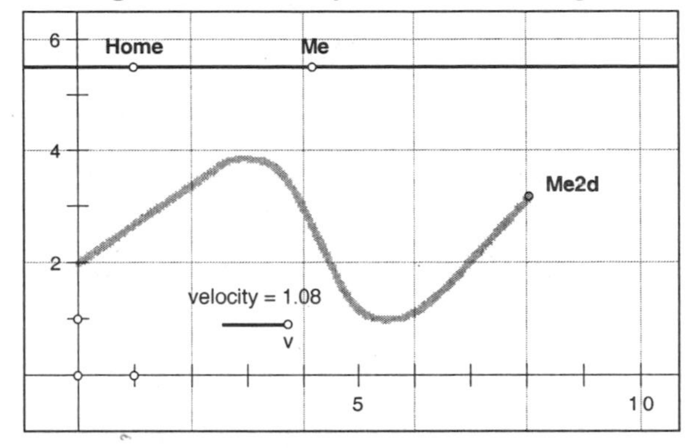

Sketch and Investigate

1. **Open** the sketch **Velocity.gsp** in the **Exploring Change** folder.

You will see a horizontal line and point *Me* that moves along it. Point *Home* represents your base point or origin. You will also see the point *Me2d*. This point represents where you are at any time. The *x*-coordinate of point *Me2d* (labeled $time_{Me2d}$) represents time, and the *y*-coordinate (labeled $position_{Me2d}$) represents your position or distance from point *Me* to point *Home*.

2. Drag point *Me2d* around the plane, getting used to the way point *Me*'s position along the line (in other words, distance from point *Home*) relates to *Me2d*'s location in the time/position plane (in other words, its coordinates).

Q1 Drag point *Me2d* horizontally. What happens to point *Me*? Explain.

Q2 You can drag point *Me2d* any way you'd like, but dragging in certain directions doesn't make sense given the way time works in our universe. How do you have to drag point *Me2d* so that it represents a physically possible motion of point *Me*?

Now we want to bring in velocity and see what effect it has.

3. Press the *Show Controls* button.

The grid in the figure is for clarification. It will not appear in the sketch.

You should see two sliders, one for velocity and one for a time interval. There is also a new point labeled *FutureMe*. This point is located one time interval away at the position you would reach *if* your velocity stayed constant. The *deltaT* slider is set at 1 and the *velocity* slider should be set at 2. So point *FutureMe* should be to the right 1 unit and up 2 units.

Q3 If you change *deltaT* to 0.5 and keep the velocity the same, what will happen to point *FutureMe*? Try it and see.

Q4 Move the *deltaT* slider to various time intervals. Does point *FutureMe* move in any particular pattern? What happens to point *Me* or point *Me2d* when you change just the time interval? Why is that?

Q5 Set *deltaT* back to 1 and now move the *velocity* slider to various values. Does point *FutureMe* move in any particular pattern? What happens to point *Me* or point *Me2d* when you change just the velocity? Why?

The *Start Motion* button will start both points moving in relationship to the set *velocity* and *time* intervals.

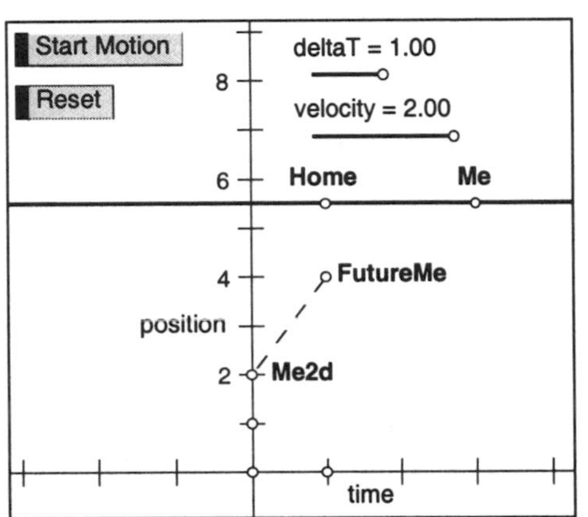

Select point *Me2d*, then choose **Trace Point** from the Display menu. You can also change the color of your selected point and trace in the Color submenu of the Display menu.

4. Press the *Reset* button to move point *Me2d* to *time* = 0.
5. Turn on tracing for point *Me2d*.
6. Set the *velocity* slider to 2 and the *deltaT* slider back to 1.

For these first trials, you won't change the *velocity* slider once your point is moving. Predict what kind of position trace you'll get if your velocity (speed and direction) stays the same. Sketch this prediction in the margin.

Q6 Press the *Start Motion* button and observe point *Me*'s motion and point *Me2d*'s corresponding time/position trace. Press the button again to stop the motion. Describe your trace. (Was it what you predicted?)

Q7 Press the *Reset* button, but do not clear your trace. Instead, change the *velocity* slider to 0.5 and make point *Me2d* a different color. Make a prediction, and then press the *Start Motion* button again. What happened this time? How are your traces different? How are they the same?

Q8 Repeat Q7, but this time set your *velocity* slider to a negative value. Any idea what will happen? Press the *Start Motion* button again. What happened this time? How are your traces different? How are they the same?

Q9 What conclusions can you reach about movement and position traces when velocity is constant over a time interval?

Q10 What are the equations for the different traces you see on your screen? What would the equation for the trace be if velocity were set to 0?

Visualizing Change: Velocity (continued)

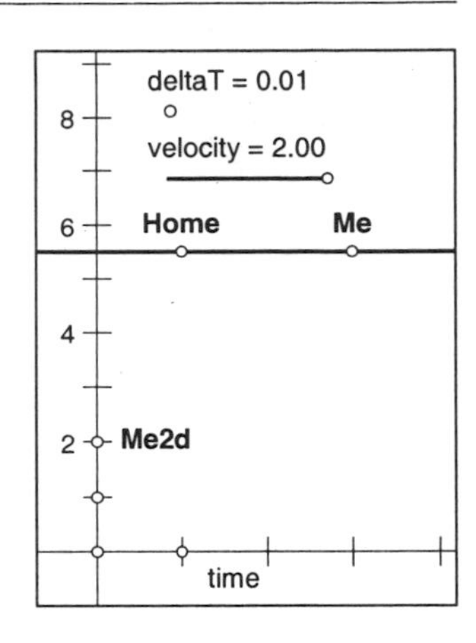

For the next set of trials, you will change the velocity of point *Me* while time is changing. The smaller the time interval, the more accurate the trace, so set *deltaT* as close to 0.1 as possible and hide point *FutureMe*. You can change the *velocity* slider to any value you wish, but try each of these suggested experiments as well. For each experiment, draw a little sketch of your trace in the margin. Remember to choose **Erase Traces** from the Display menu and press the *Reset* button when you want to start over.

To hide a point, select the point and then choose **Hide Plotted Point** from the Display menu.

A. Start with the velocity at a positive value. Increase the velocity, and then decrease the velocity, but keep it positive throughout the experiment.

B. Start with the velocity at a negative value. Increase the velocity, and then decrease the velocity, but keep it negative throughout the experiment. (Remember that –2 to –1 is an increase!)

C. Start with *velocity* > 2. Decrease the velocity, and then increase it. Again, keep the velocity positive throughout.

D. Start with –1 < *velocity* < 0. Decrease the velocity, and then increase it, but again, keep the velocity negative throughout.

E. Start with a positive velocity and decrease to a negative value. Then increase the velocity again until you get to 0. Stay at 0 for a while and then increase the velocity again.

Q11 How are the traces in A and B similar? How are they different? What happens to the position trace when you switch from increasing the velocity to decreasing it?

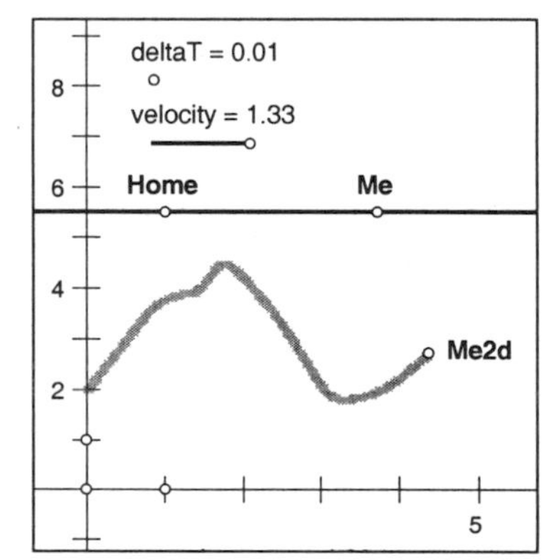

Q12 How are the traces in C and D similar? How are they different? What happens to the position trace when you switch from decreasing the velocity to increasing it?

Q13 How are the traces in A and C similar? How are they different? What about B and D?

Q14 What happened when you changed the velocity from positive to negative? From negative to positive? What happened when you stayed at *velocity* = 0?

Q15 For each of the following, describe the position trace that you would get. Then check your answer using the *velocity* slider.

a. positive and increasing velocity

b. negative and increasing velocity

c. positive and decreasing velocity

d. negative and decreasing velocity

Explore More

Go to page 2 of the sketch. Press the *Show Path1* button. Using your answers from Q15 for reference, make a trace trying to match the path as closely as you can. During which part of your trace did you have to go the fastest? When did you move the slowest?

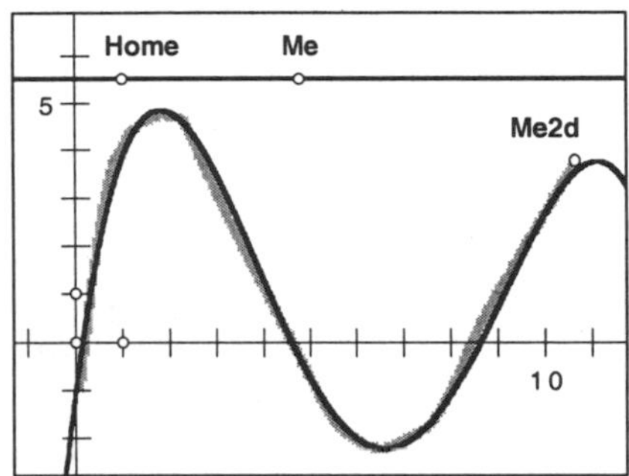

Hide Path 1 and press the *Show Path2* button. Again, try to match the path as closely as you can.

What is different about Path 2? Which one was easier to trace? Is it possible to trace Path 2's corners?

Can You Predict the Trace?

Name(s): ______________________

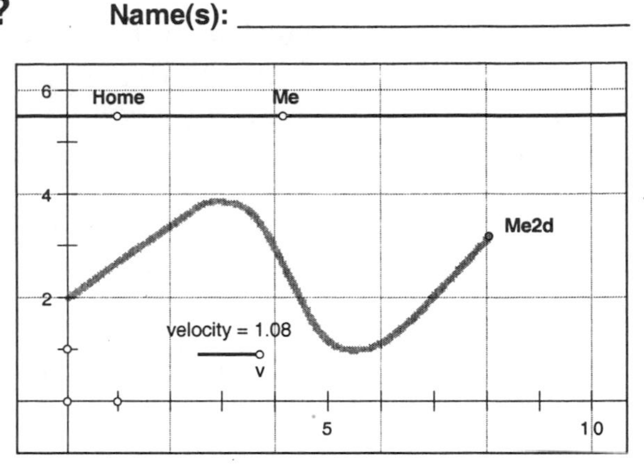

In the previous activity you created a position plot by controlling the velocity of the object. You saw that what you did with the velocity determined what the position trace looked like. Could you look at a position plot and predict the velocity plot? In this activity, you will trace position and velocity at the same time and use what you discover to predict what one graph would look like given the other.

Sketch and Investigate

1. **Open** the sketch **RemoteControl.gsp** in the **Exploring Change** folder. (To see how this sketch works, review pages 7–9.)
2. Select the measurements of time and velocity, in that order, and choose **Plot as (x, y)** from the Graph menu.
3. Turn on tracing for this new point and label it *MyVelocity.*
4. Press the *Move* button, but don't change the *velocity* slider. Just watch.

To stop the experiment, press the *Move* button again.

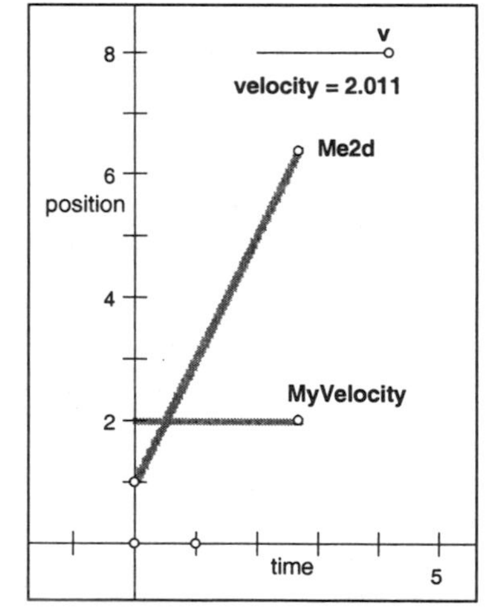

You should see two traces created—one for position and one for velocity. Since the velocity is not changing, its trace is a horizontal line.

Q1 The position trace is linear; what is its slope? Have you seen this value anywhere else?

5. Press the *Reset* button and choose **Erase Traces** from the Display menu to return point *MyVelocity* and point *Me2d* to their original locations so you can run a new experiment.

This time, change the velocity after you press the *Move* button. Increase or decrease the velocity as you wish, but try each of these suggested experiments as well. For each experiment, draw a little sketch of your trace in the margin. Remember to press the *Reset* button and choose **Erase Traces** from the Display menu when you want to start over.

A. Start with the velocity at a positive value. Increase the velocity, and then decrease the velocity, but keep it positive throughout the experiment.

B. Start with the velocity at a negative value. Increase the velocity, and then decrease the velocity, but keep it negative throughout the experiment. (Remember that –2 to –1 is an increase!)

C. Start with *velocity* > 2. Decrease the velocity, and then increase it—again keep the velocity positive throughout.

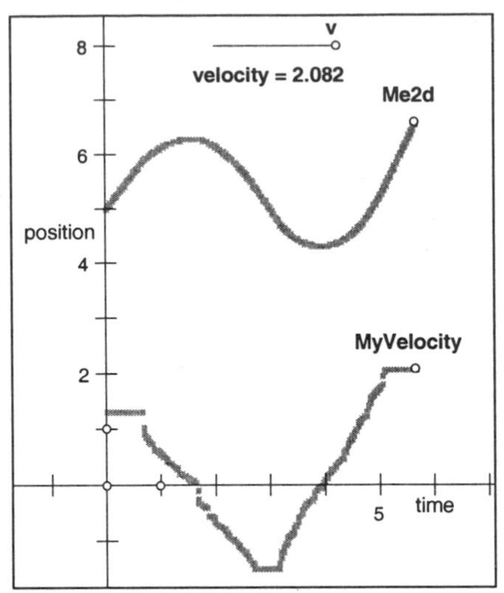

D. Start with –1 < *velocity* < 0. Decrease the velocity, and then increase it—but again keep the velocity negative throughout.

E. Start with a positive velocity and decrease to a negative value. Then increase the velocity again until you get to 0. Stay at 0 for a while and then increase the velocity again.

Q2 When your position trace is increasing (going up), what can you say about your velocity trace? (Or what must you do with the *velocity* slider to make the position trace keep going up?)

Q3 When your position trace is decreasing (going down), what can you say about your velocity trace? (Or what must you do with the *velocity* slider to make the position trace keep going down?)

Q4 When your position trace hits a maximum or minimum, what happens to the velocity trace? (Or what must you do to make your position trace change direction?)

Q5 When your position trace is *concave up* (curved up), what can you say about your velocity trace? (Or what must you do with the *velocity* slider to make the position trace curve upward?)

Q6 When your position trace is *concave down* (curved down), what can you say about your velocity trace? (Or what must you do with the *velocity* slider to make the position trace curve downward?)

These questions suggest some connections between the plot of the position of a point and a plot of the velocity of the point. On the second page of the document, you can try predicting the trace of velocity from the trace of position.

6. Press the *Predict Velocity Trace* button to go to page 2.

On this page, the velocity slider will be controlled by a hidden function. Instead of adjusting the velocity slider, you can watch the velocity slider and the position trace and make a prediction about what the velocity trace would look like.

7. Press the *Move* button and observe the position trace and the velocity slider. The plotted point that tracks the velocity is hidden.

8. Make a sketch in the margin of what you think the velocity trace will look like, using your answers to the questions above. After you make your prediction, press the *Reset* button, then the *Show Point MyVelocity* button, and then the *Move* button to see the trace of position and velocity at the same time.

Q7 How does the velocity trace compare to your prediction in step 8?

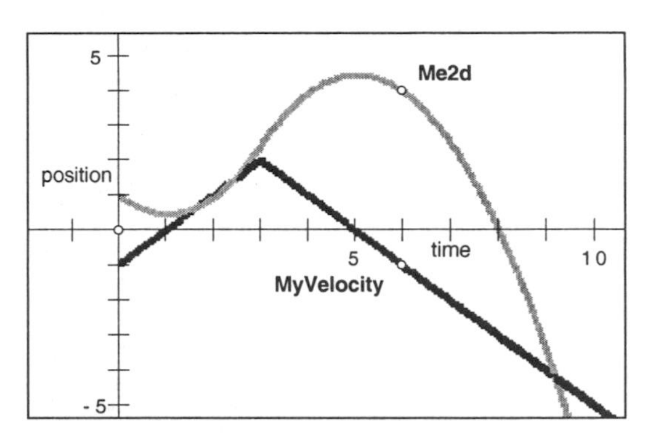

9. To see another example, press the *Next* button and repeat steps 7–8. You can practice on any of the eight examples on pages 2–9 in the document. When you can accurately predict the velocity trace from the given position trace for one example, go on to the next by pressing the *Next* button.

Next, you will try to predict the position trace given the velocity trace. First, consider the questions below. (You can use your answers to Q2–Q6 and page 1 of your document to answer these questions.)

Q8 When your velocity is positive, what can you say about your position trace?

Q9 When your velocity is negative, what can you say about your position trace?

Q10 When your velocity trace hits a maximum or minimum, what happens to the position trace? (What happens to the position trace when the velocity slider changes direction?)

Q11 When your velocity trace is increasing, what can you say about your position trace? (What happens to the position trace when the velocity slider moves to the right?)

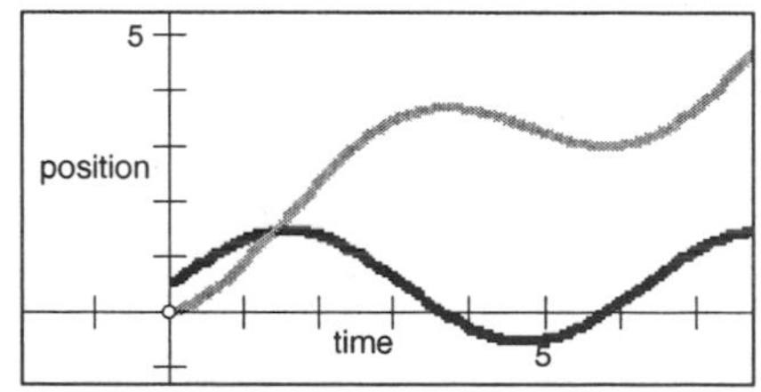

Q12 When your velocity trace is decreasing, what can you say about your position trace? (What happens to the position trace when the velocity slider moves to the left?)

These questions suggest some connections between the plot of the velocity of a point and a plot of the position of the point.

The *Predict Position Trace* button is on page 1 of your document.

10. Press the *Predict Position Trace* button to go to page 10.

Point *Me2d* is hidden on this page. Here you'll use the velocity trace and the position of point *Me* on the line in relation to point *Home* to make a prediction about what the position trace would look like.

11. Press the *Move* button and observe the velocity trace and the position of point *Me*.

12. Make a sketch in the margin of what you think the position trace will look like, using your answers to the questions above. After you make your sketch, press the *Reset* button, then the *Show Point Me2d* button, and then the *Move* button to see the trace of position and velocity at the same time.

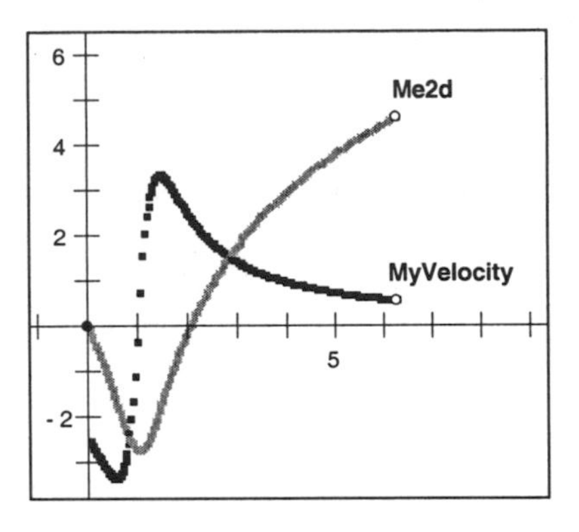

Q13 How does the position trace compare to your prediction in step 12?

13. To see another example, press the *Next* button and repeat steps 11–12. You can practice on any of the six examples on pages 11–16 in the document. When you can accurately predict the position trace from the given velocity trace for one example, go on to the next by pressing the *Next* button.

Explore More

Point *R* sets the starting location for point *Me2d* in the predicting position trace pages. Try moving this point, pressing *Reset*, and running the experiment again. (The velocity function will remain unchanged.) Compare the position traces for two different initial positions. What do you notice? Why do you think this happens?

Catching the Point

Name(s): ____________________

In the last two activities, you looked at position plots or motion by moving a point—either by dragging the point to various places or by controlling its velocity with a slider. You saw that there was some kind of relationship between the point's position and its velocity. What if the point moves according to some pattern that is not in your control? Is there still a relationship, and what does all this have to do with calculus? In this activity, you'll start to answer both questions.

Sketch and Investigate

1. **Open** the sketch **Motion.gsp** in the **Exploring Change** folder. Point Q moves on the horizontal line according to a pattern—a function that gives its directed distance from point *Home* at a specific time.

To measure the coordinates, choose **Abscissa(x)** from the Measure menu and then choose **Ordinate(y).** Then to calculate the distance, choose **Calculate** from the Measure menu, and click on each measurement to enter it into the calculator.

2. Measure the x-coordinates of point *Home* and point Q and calculate the distance from point *Home* to point Q, or $time_Q - time_{Home}$. Label this measurement *position.*

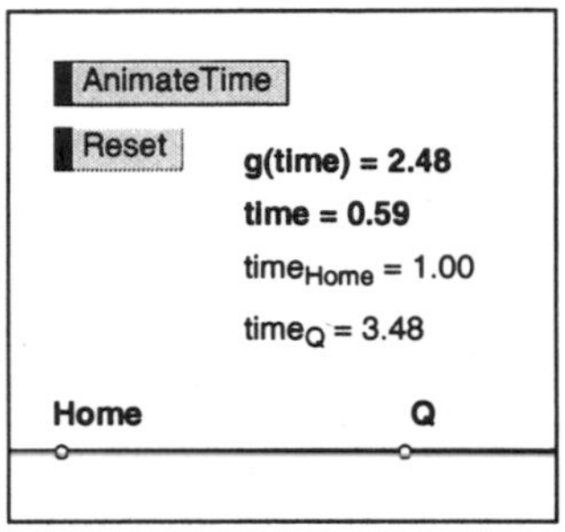

AnimateTime

Reset

g(time) = 2.48

time = 0.59

$time_{Home}$ = 1.00

$time_Q$ = 3.48

Home Q

Now press the *Animate Time* button—watch how the point moves and how fast it moves. Do this a few times—check that the *position* measurement and the function value match at *all* times, watch how the velocity changes, and try to figure out when point Q changes direction. You can stop the animation at various times by pressing the *Animate Time* button again. This is most helpful for figuring out when the point changes direction.

Q1 Describe point Q's path. For what values of time does it go to the right? Go to the left? Stop?

Q2 When does it seem to move the fastest? The slowest?

It's difficult to figure out these kind of questions when your movement is limited to one dimension—especially if you can't stop time. So creating a position plot of *position* versus *time* is the next step.

3. Select measurements *time* and *position*, in that order, and choose **Plot as (x, y)** from the Graph menu. Label this point P.

To turn on tracing, select point P, then go to the Display menu and choose **Trace Plotted Point.** You can also change the color of your selected point and trace by choosing **Color** in the Display menu.

4. Turn on tracing for point P and color it a bright color.
5. Press the *Animate Time* button. After point P traces one cycle, press it again to stop. Does point P change direction at the same times as point Q? Do the two points seem to go at the same speed?

Q3 Describe point P's path—when does point P go up? Go down? Change direction? Is there a relationship between your answers for point Q and point P?

Q4 Describe point P's motion. When does point P move quickly—either up or down? When does it slow down?

To figure out if there is a pattern or relationship between the velocity and the position of point Q, you are going to make a model by using an independent point, point *Me2d,* and some sliders. Your goal is to construct a point, *FutureMe,* that will help you control point *Me2d*'s path and velocity. Then you'll try to match point P's path.

To construct a point, choose the **Point** tool, and then click anywhere in the sketch. To label the point, double-click on it with the **Text** tool.

6. Construct a point anywhere in your sketch using the **Point** tool. Label this point *Me2d* and measure its x- and y-coordinates. Label these measurements *myTime* and *myPosition*.

7. Press the *Show Sliders* button to show a time interval slider, labeled *deltaT,* and a slider for velocity.

Right now, you're sitting at the point (*myTime, myPosition*). If some time passes and you are moving with a constant velocity, where will point *FutureMe* be located? Use the diagram to help.

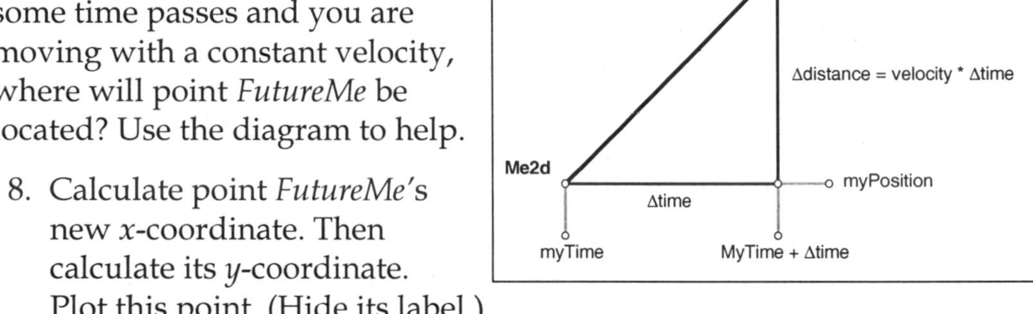

Choose **Calculate** from the Measure menu. Then click on each measurement to enter it into the calculator.

8. Calculate point *FutureMe*'s new x-coordinate. Then calculate its y-coordinate. Plot this point. (Hide its label.)

To construct a line segment, select each point then choose **Segment** from the Construct menu.

9. Construct the line segment between point *Me2d* and point *FutureMe.* This line segment represents your path, so color it a different color than point P.

Q5 If you shorten or lengthen your time interval, *deltaT,* how does that affect your path?

Q6 If you shorten or lengthen your velocity slider, how does that affect your path?

Be sure to deselect all objects first.

10. To start moving, select just point *Me2d* and then select point *FutureMe.* Choose **Edit | Action Buttons | Movement.** Select **slow** for the speed and label the button *Move Me.*

By pressing the *Move Me* button, you can experiment with using different time intervals and velocities to move your path.

Press the Esc key once to stop. Then, to start over, drag point *Me2d* back to a starting point and choose **Display | Erase Traces.** Then press the *Move Me* button.

11. Set $deltaT \approx 2$ and $velocity \approx 0.5$. Drag point *Me2d* to (0, 2) to start. Turn on tracing for your segment. Press the *Move Me* button. Keep your *deltaT* constant but experiment with your *velocity* slider.

Q7 What happens to your path trace if you make a large change in your velocity?

12. Set *deltaT* ≈ 0.1 and *velocity* ≈ 1. Drag point *Me2d* to (0, 2) to start. Erase all traces. Press the *Move Me* button. Again keep your *deltaT* constant but experiment with your *velocity* slider.

Q8 Now what happens to your path trace if you make a large change in your velocity?

Because the point's construction assumed a constant velocity over a time interval, you'll want to make *deltaT* small so that you can change your velocity frequently. Your goal in the next part will be to use your *velocity* slider to trace a path as close to point *P*'s path as possible.

13. Press the *Show P's Path* button and set *deltaT* ≈ 0.50.

14. For now, turn off tracing for the segment by selecting the segment, and then deselecting **Trace Segment** on the Display menu. Erase all traces.

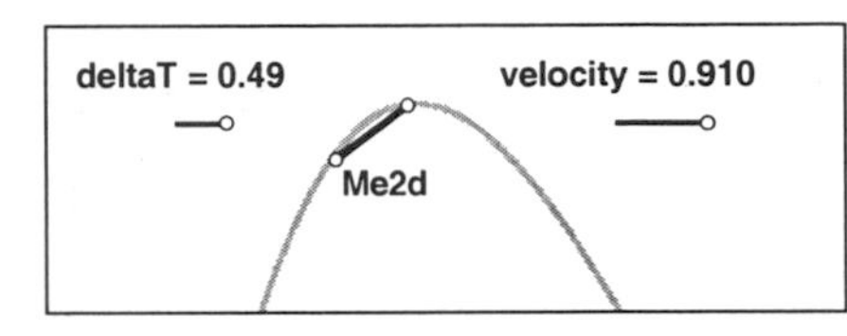

15. Drag point *Me2d* to anywhere on point *P*'s path and adjust your velocity until point *FutureMe* is also on the path. (This is roughly point *Q*'s velocity—note that it doesn't match the path exactly.)

16. Press the *Move Me* button and try to stay on point *P*'s path by using your *velocity* slider.

17. Try several more times, each time decreasing your time interval by 0.10, but don't go below 0.

Q9 Does your accuracy improve as *deltaT* gets closer to 0? Can you stay on point *P*'s path the whole time?

Q10 What must you do with your *velocity* slider in order to make your path go up? Go down? Change direction?

Once you can stay on point *P*'s path most of the time, you're ready to model the velocity. You'll do this by making a new trace.

deltaT should still be close to zero.

18. Press the *Reset* button, then drag your point *Me2d* back to point *P*'s starting point, (0, –1). Line up point *FutureMe* with point *P*'s path by increasing or decreasing your *velocity* slider.

19. Once you are lined up on point *P*'s path, select *myTime* and *velocity* in that order and choose **Plot as (x, y)** from the Graph menu. This will track the velocity of point *Me2d*.

You may need to scroll up your window in order to find your velocity point. After you label it and turn on tracing, scroll back down.

20. Label this point *V*, turn on tracing, and color point *V* a new bright color.

21. Press the *Move Me* button and make sure you stay on point *P*'s path as you did before. (You can start over if you "fall off.")

22. Stop and examine your new trace. Do the patterns between the position and velocity functions hold from the previous activity? (See the Activity Notes if you didn't do the previous activity.)

So how does calculus come in? The little segment path you traced really used the average velocity in its construction because we assumed the velocity stayed constant over a time interval. In calculus we ask, "What is the velocity *now*—in this instant or at this moment?" This is called *instantaneous velocity*. You might think we can't use the average velocity at all to answer this question because the average velocity is defined as the change in distance divided by the change in time, and if we're talking about an instant, the change in time is 0. That would give us a 0 in the denominator—a major problem! One way to start thinking about this is to ask the question, "What happens if I let *deltaT* get really close to 0 or *go to* 0, but never really get there—what will I see?" Calculus answers this question using a concept called limits.

Press the *Show Q's Velocity* button. This is point Q's instantaneous velocity as a function of time. How close did your velocity trace come to matching this? Beside the jerkiness of a trace, if you did a fairly good job of staying on point P's path and had a very tiny *deltaT*, your velocity trace should be fairly close to the plotted velocity.

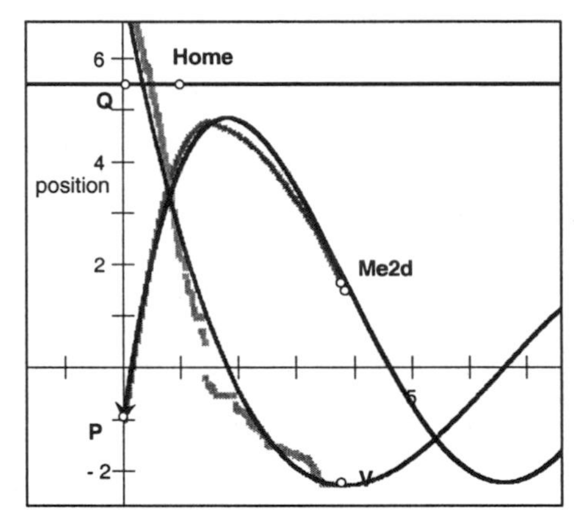

Explore More

Press the *Hide Q's Velocity* button. Select point P's path. Choose **Properties** from the Edit menu. Go to the Object panel and choose Function g from the Parents menu. Deselect the Hidden box and click OK. The function highlighted in the sketch is the one that moves point P and point Q. Double-click on this function and change it to $g(x) = \text{abs}(x^2 - 12x + 32)$.

Try to trace this function's path. How is this path different from the previous function's path? How does that affect the velocity?

Once you have a velocity trace that you are satisfied with, press the *Show Q's Velocity* button. Select the velocity plot, then choose **Properties** from the Edit menu. On the Plot panel, change **continuous** to **discrete.** Click OK. Compare your trace of the velocity to the plot. How did you do?

Plotting Average Rates of Change

Name(s): ______________________

In this activity, you will examine two ways of describing average rates of change. In the process, you will be introduced to two new tools and investigate how to apply them to various functions as you create an average rate of change plot. (If you would like to build these tools yourself, go to **Tool Tips.doc** in the **Tools** folder.)

Sketch and Investigate

1. **Open** the document **Step.gsp** in the **Exploring Change** folder.

2. Plot two points anywhere in the coordinate plane using the **Point** tool. Label them point *A* and point *B* with your **Text** tool.

For any two points $A\ (x_A, y_A)$ and $B\ (x_B, y_B)$, the average rate of change of y with respect to x is expressed by the ratio $\frac{y_B - y_A}{x_B - x_A}$.

The **Custom** tools icon is at the bottom of the toolbar.

3. Choose **Average Rate** from **Custom** tools. Click on point *A* and then point *B*.

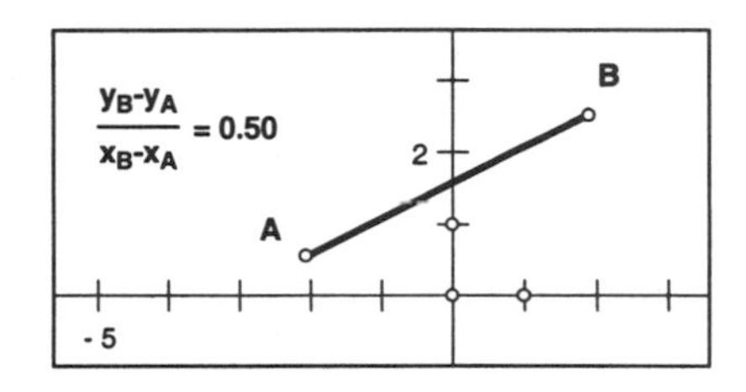

This tool calculates the average rate of change from point *A* to point *B*. Drag points *A* and *B* around to observe the changes in the ratio.

Q1 When is the average rate of change from point *A* to point *B* positive? When is it negative?

Q2 Does point *A* need to be on a particular side of point *B* in order to have a positive rate of change? Why or why not?

Be sure not to select points *A* and *B* here.

4. Select segment *AB*. Choose **Slope** from the Measure menu.

Q3 Drag point *A* and point *B* around to observe the changes in both measurements. Why do they always match? (Move point *B* back to the right of point *A* when you're done.)

Now that you have found the rate of change between two points, you will look at a series of points on a function plot.

To enter *a* and *b* into the calculator, click on their measurements.

5. Using the measurements provided, plot the function $f(x) = ax + b$ by choosing **Plot New Function** from the Graph menu.

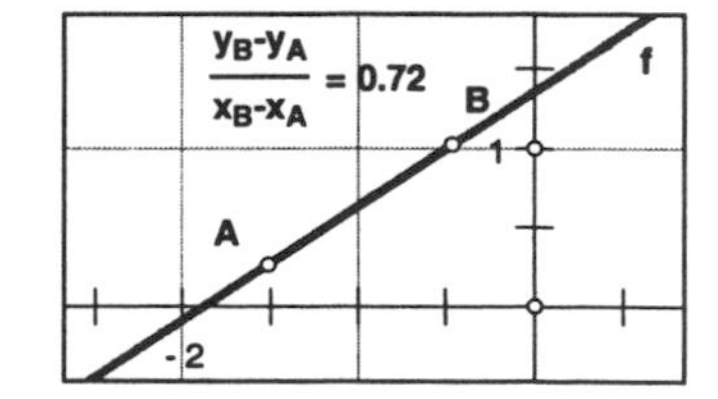

6. Click on an empty spot in the sketch to deselect $f(x)$.

7. Select point *A* and then the plot of $f(x)$—not the measurement. Then choose **Merge Point to Function Plot** from the Edit menu. Merge point *B* to the function plot as well.

8. Choose **Average Rate** from **Custom** tools, click on point *B*, and then click on the function plot to the right of point *B* to make a new point.

9. Calculate the average rate of change on the next adjacent interval. (With the **Average Rate** tool, click on the point you made in step 8, and then on a new point on the function to the right of that point. Interval lengths can be different.)

Q4 Why are all your average rate measurements equal to each other? Does moving any of the points you created change that?

If you can't see your function plot, drag the slider for *c* until it comes into view.

10. Double-click on the expression for the function $f(x)$. Change $ax + b$ to $ax^2 + bx + c$.

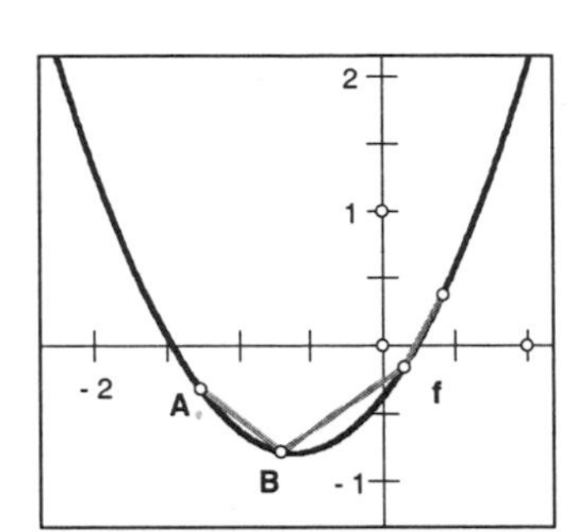

Q5 Are all the average rate measurements still equal to each other?

Q6 How can you determine from the function plot where the average rate of change between two points on that function will be positive, negative, or 0?

As you saw in Q3, the average rate of change between two points can be seen graphically by the slope of a segment between those points. Can slope be plotted as a function as well? Slope is defined over an interval, so the question is, how do you plot a quantity defined over an interval? One way is with a step function.

11. Go to page 2 of the document.

12. Choose **Plot Slope** from **Custom** tools, and then click on points *A* and *B* in that order. This tool plots the slope between two points *A* and *B* on the interval $[x_A, x_B]$ as a "step"—or horizontal segment.

Q7 Experiment with this "step" by dragging either point *A* or point *B* around in the plane. When does the step have a negative *y*-value? When does it have a positive *y*-value? When is it 0?

Go back to page 1 to use the **Plot Slope** tool on the points you created.

When creating a new step, click first on the point you just created for the previous step. You will be clicking twice on each endpoint where two segments meet.

13. With the **Plot Slope** tool, click on *both* endpoints of each segment you created on $f(x)$ in steps 8–9. Then use your tool to create more points as you did in step 9, until you have a total of eight "steps."

Q8 Your "steps" make up a new relation that plots the average rate of change for each interval that you created on $f(x)$. With the **Arrow** tool, press the *Case a = 0* button to set $a = 0$. Explain why this new function plot of "steps" is a horizontal line.

Q9 Experiment with the slider for *c*. Describe what happens to your step function as a result and explain why.

Q10 Experiment with the slider for b. Describe what happens to your step function as a result and explain why.

Again, if you can't see your function plot, drag the slider for *c* until it comes into view.

14. Make your measurement for a greater than 0.

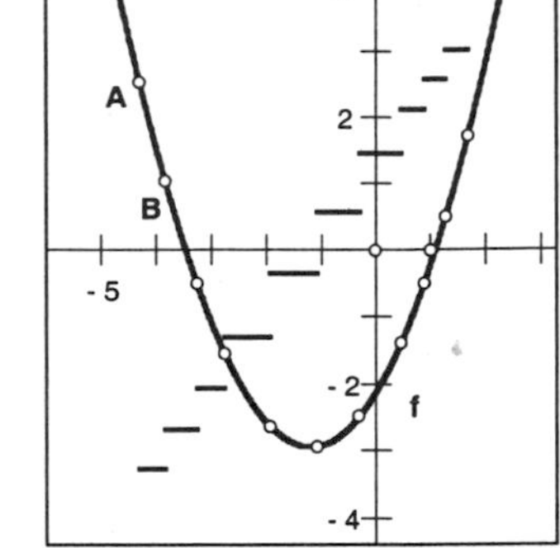

Q11 In the figure to the right, the steps go up from left to right. Why? How can you adjust your function so that they go down?

Q12 Predict what will happen if you again adjust the value for c. Explain why this happens. Check your answer.

You now have a new function that plots the average rate of change for each interval that you created on the function $f(x)$.

Q13 Go to page 3 and predict what the step function will look like for that cubic. How can you determine from the function plot when the rate of change step function will be positive (above the x-axis), negative (below the x-axis), or zero?

Press the *Show Steps* button. How did you do?

Explore More

On page 4 of the document you will find the plot of a function, with two points A and B on the function. Here the slider labeled h controls the distance between the x-coordinates of the two points.

1. Using the **Plot Slope** tool from **Custom** tools, click on points A and B to create a "step" of the rate of change on the interval from point A to point B.
2. With the segment still selected, choose **Trace Segment** from the Display menu.
3. When you press the *MoveA→B* button, the two points will move to the next adjacent interval. Press the button a few more times to create a trace of the step function.

To edit the speed, double-click on the *MoveA→B* button with your **Text** tool and go to the Move panel.

4. Adjust your slider so that the measurement $h \approx 0.1$ and edit the *MoveA→B* button so that the points move at **medium** speed.
5. Drag point A to the left side of the window and choose **Erase Traces** from the Display menu. Then press the *MoveA→B* button. Press it again to stop.

Q1 Try this for a number of functions, and record your observations. Can you predict what the trace of the average rate of change "step" will look like, given the plot of the function?

Going the Distance

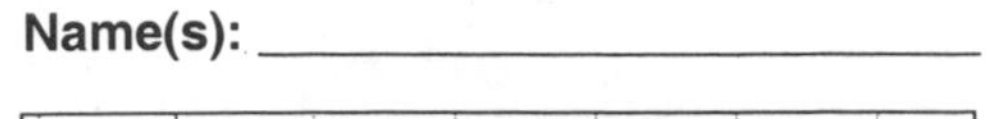

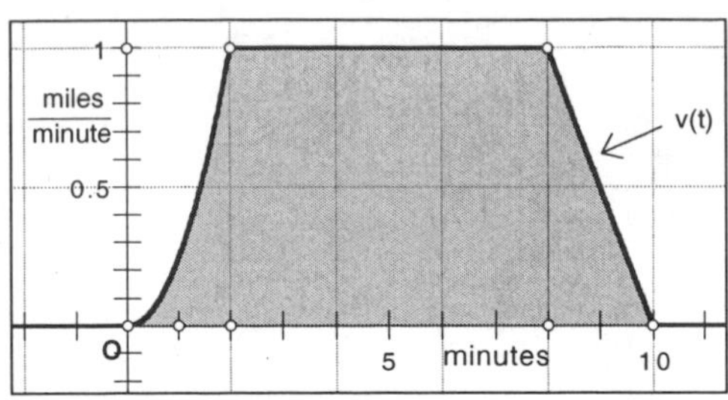

When you drive a car, you can always find your velocity just by looking at your speedometer. Can you figure out your distance from that? Yes, but only if you are driving at a constant speed. But that is nearly impossible, and you still have to accelerate when you start and decelerate when you stop. At these times, your velocity isn't constant at all. The sketch above depicts a possible velocity function that fits these conditions. (Note that here we're assuming you can drive at a constant velocity—we'll pretend you have perfect cruise control.) Given this velocity function, how could you find your total distance? Why is the region below the function shaded in? If you've guessed that the two questions are probably related, you're right. Let's find out why.

Sketch and Investigate

1. **Open** the sketch **Area1.gsp** in the **Exploring Change** folder. Here you have the velocity function shown above, without the shading.

Q1 For how many minutes is your velocity constant? What is your velocity for those minutes?

Q2 Because your velocity is constant for this segment of your trip, you can figure out your total distance for this interval of time. What is this distance?

Be sure to select points *A*, *B*, *C*, and *D* in that order.

2. Construct the interior of rectangle *ABCD* by selecting the vertices of the rectangle and choosing **Construct | Quadrilateral Interior.**

Sketchpad's Area command uses inches, which is not relevant here, so calculate this by hand.

Q3 What is the area of rectangle *ABCD*? Include units in your answer!

Q4 Are the answers to Q2 and Q3 related in any way? Can you jump to any conclusions at this point?

Look at just the units of the axes—the horizontal axis is in minutes and the vertical axis is in miles per minute. So the units for any area involving a base and a height will be

$$base \cdot height = minute \cdot \frac{miles}{minute} = miles$$

But miles is purely a distance measurement, so in this case, the area under the velocity curve from $x = 2$ to $x = 8$ is the total distance traveled in that time. Is that always true? Maybe. Let's look at another example and see whether this conjecture holds. Consider the velocity function from $x = 8$ to $x = 10$. Here, instead of being constant, the velocity is a linear function that goes from 1 mile/min at $x = 8$ to 0 mile/min at $x = 10$. Here, too, you can calculate your total distance using your average velocity. (Why?)

Q5 What is the average velocity on the interval [8, 10]? Using this, what is your total distance from $x = 8$ to $x = 10$?

Color this triangle a bright new color from the Color submenu of the Display menu.

3. Construct the interior of triangle *BCE* by selecting the vertices of the triangle and choosing **Triangle Interior** from the Construct menu.

Q6 What is the area of triangle *BCE*? Include units in your answer.

Again, the area under this velocity curve from $x = 8$ to $x = 10$ is the total distance traveled in that time. Is this true all the time? Actually, it is, but the proof of that is a bit off in the future. For now, what's important is that the area under a velocity curve over an interval gives the total distance traveled in that interval. So you can find your total distance if you can calculate an area. Sounds simple, right? Then how are you going to calculate the area under the velocity curve from $x = 0$ to $x = 2$?

Q7 What familiar polygon most closely approximates this area?

Q8 Find an approximation for the total distance by finding the area of the polygon you chose in Q7. (Save this answer.)

4. Construct the interior of your polygon.

Q9 Comparing the interior you constructed in step 4 and the actual region under the curve, is your approximation in Q8 too big or too small?

5. Select the interior of the polygon and choose **Display | Hide Triangle.**

To get a closer approximation, you could divide the interval [0, 2] into two sub-intervals and use two different polygons.

6. Select the unit point at $x = 1$ and measure its x-coordinate by choosing **Abscissa(x)** from the Measure menu.

7. Press the *Show Function* button and calculate $f(x_U)$ by choosing **Calculate** from the Measure menu. Click on the expression for f and the measure for x_U to enter them into the calculator.

Double-click on the point with the **Text** tool to change a label.

8. Select x_U and $f(x_U)$, in that order, and choose **Plot as (x, y)** from the Graph menu. Label this point *G*.

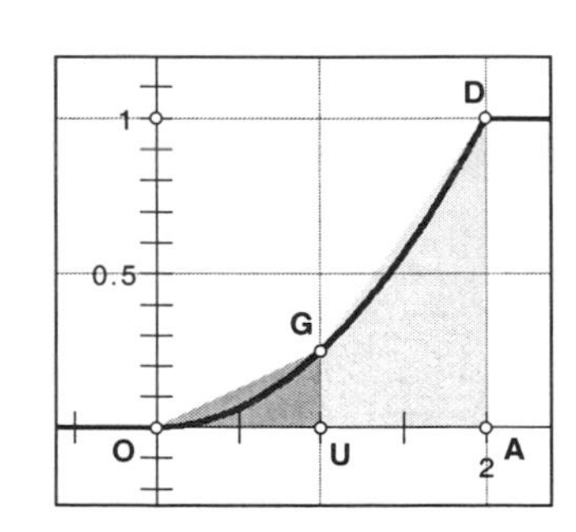

9. Choose any of the light color tools from **Custom** tools and click on points *O*, *U*, and *G*, and then on point *O* again to construct the interior of triangle *OUG*.

10. Choose a different color tool from **Custom** tools and click on points *U*, *A*, *D*, and *G* to construct the interior of the trapezoid *UADG*.

Q10 What are the areas of triangle *OUG* and trapezoid *UADG*? (Recall that the area of a trapezoid is $A = 0.5 \cdot base \cdot (height_1 + height_2)$.)

Q11 What is your total area approximation? When you compare your shaded regions with the area under the curve, is this total area too big or too small? (Save this answer.)

Q12 Did your approximation improve significantly? (Is a lot less of the shaded region outside of the curve now?)

Press the *Show Area Tools* button. On the *time* axis, there are three new points—*start, finish,* and *P*. Points *P* and *start* should be at the origin. Point *P* will sweep out the actual area under the curve from point *start* to point *finish,* so point *finish* is as close as possible to $x = 2$. The measurement *Area* should be 0 because point *P* has not moved yet.

11. Press the *Move P* button to calculate the actual area under the *velocity* curve and to shade in that region.

Q13 What is the actual area under the curve from $x = 0$ to $x = 2$?

Q14 Calculate your approximation errors from Q8 and Q11. Was your second approximation a significant improvement?

One way to tell whether your second approximation is a significant improvement is to calculate the percentage errors—divide your errors in Q14 by the actual area. You should have a drop from about 50% error to about 12% error. That is significant!

Q15 Using your answers to Q3, Q6, and Q13, what is the total area under the velocity curve from $x = 0$ to $x = 10$? This is the total distance you traveled in those 10 minutes!

Exploration 1

To get an even better approximation, you can do the same procedure again. Divide the region on [0, 2] into fourths, using one triangle and three trapezoids.

1. Using the **Point** tool, construct new points at $x = 0.5$ and $x = 1.5$ or as close to those values as possible. Label these points *P* and *Q*.

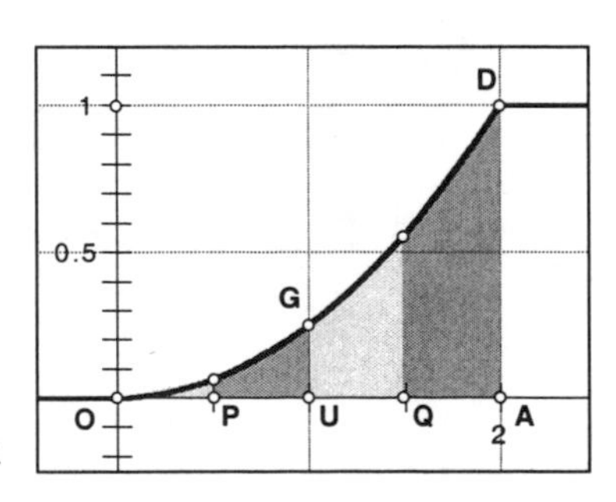

2. Construct the y-values on the function for these two points as you did in step 6 above. Then plot the points on the function at $x = 0.5$ and $x = 1.5$ as you did in step 8.

3. Construct the interior of the triangle on the interval [0, 0.5], using a color tool as you did in step 9. Calculate its area—this time with your measurements x_P and $f(x_P)$ and the calculator.

4. Construct the interiors of the trapezoids over the intervals [0.5, 1], [1, 1.5], and [1.5, 2]. Use a different color tool for each one. Calculate the areas of the three trapezoids using your measurements.

5. Calculate a new approximation for the area under the curve from $x = 0$ to $x = 2$ by making a new measurement that sums the calculations you did in steps 3 and 4 above.

Q1 What is your new approximation for the area under the curve from $x = 0$ to $x = 2$?

Q2 What is the error for your new approximation? What is the percentage error? Was this new approximation a significant improvement?

To get an even better approximation, you can do the same procedure again. Divide the region on [0, 2] into eighths, using one triangle and seven trapezoids. Then do it again by dividing your region into sixteenths and so on. This can quickly become tedious, so in the activity "The Trapezoid Tool," you will make a tool that makes trapezoids and measures their area.

Exploration 2

Point U is fixed because it is the unit point (trying moving it), but point P and point Q are not fixed. You constructed point P and point Q instead of plotting them exactly, so you can move them along the x-axis and see whether you can come up with a better approximation—without constructing more trapezoids.

1. Move point P back and forth in the interval [0, 1], and look at the percentage error that you calculated back in Q2 of Exploration 1—or you can look at the shaded regions above your curve because they also represent your error.

Q1 For approximately what x-value is the error the smallest? (If you are looking at the shading, remember you want the shading to be smallest above all parts of the curve between $x = 0$ and $x = 2$.)

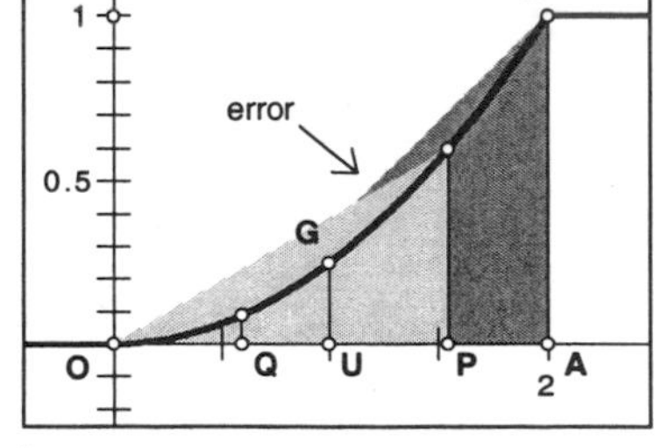

2. Move point Q back and forth in the interval [1, 2], and look at your percentage error from Q2 of Exploration 1.

Q2 For approximately what x-value is the error the smallest?

What Do You Expect?

Name(s): ____________________

Limits are one of the fundamental building blocks of calculus. So what is a limit exactly and how do you find it? This activity will help you answer these questions.

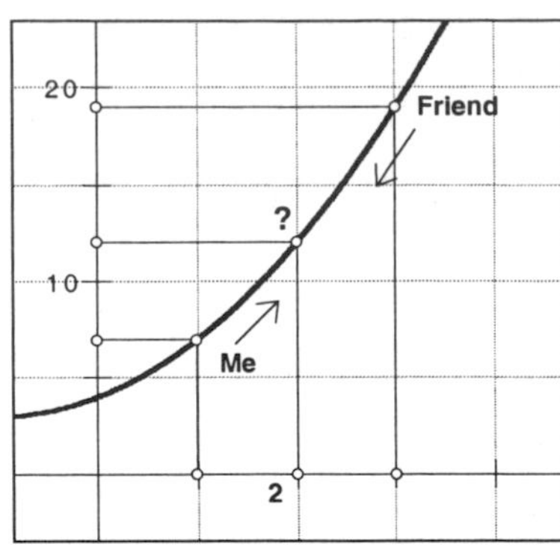

Sketch and Investigate

Imagine that you're hiking on a trail and it conveniently happens to be curved exactly like the function $q(x) = \dfrac{x^3 - 8}{x - 2}$, but you don't know that. All you have is an interactive map, similar to the picture at right.

1. **Open** the sketch **Limits.gsp** in the **Exploring Change** folder for your interactive map.

You can also choose **Abscissa(x)** from the Measure menu and then choose **Ordinate(y).**

2. Measure your coordinates and your friend's coordinates by selecting point *Me* and point *Friend*, and then choosing **Coordinates** from the Measure menu.

Now, you're hiking along and you *expect* the trail to be smooth—no potholes or drop-offs. Unfortunately, there's a huge pothole ahead at $x = 2$, but it's not marked on your map, so you don't know about it. What you see and *expect* is to be able to meet your friend, who's hiking down the trail from the other direction to meet you, at the point $x = 2$.

3. Drag point *Me* until you are as close as possible to $x = 2$, but keep your x-coordinate less than or equal to 2. Do the same with point *Friend,* but make sure its x-coordinate stays greater than or equal to 2.

Q1 How close to $x = 2$ can you get with each point? What y-coordinate did you get for point *Me*? What about for point *Friend*?

Q2 Based on your points and their movement, what do you *expect* the y-coordinate to be at the point $x = 2$?

What you *expect* to see is called a *limit*—actually what you expect to see is called a *left-hand limit* and what your friend expects to see is called a *right-hand limit*. Whether reality agrees with your expectations or not is a different concept altogether, and does not affect your limit.

Q3 What is the left-hand limit at $x = 2$? What is the right-hand limit?

Q4 Do both you and your friend expect to see the same y-value?

We say that a limit exists if the left-hand limit agrees with the right-hand limit. In this case, you should have gotten $y = 12$ for both parts of Q3. (If you didn't, go back and try again.) What really happens at $x = 2$ is the y-value of the function at $x = 2$, or $q(2)$.

4. Calculate the y-value of the function at $x = 2$ by choosing **Calculate** from the Measure menu. Then select the expression for $q(x)$ to enter it into your calculator, enter 2, and close the parentheses.

You should have gotten undefined for $q(2)$. So here reality (the function) does not exist, but that still doesn't affect what you *expect* to see. You still *expect* to see $y = 12$. The key to success with limits is to remember that limits are expectations only—not what really happens!

You may think, "Well, maybe my expectations (or limit) would be different if I zoomed in." Press the *Show Zoom Tools* button. The sliders, *x-scale* and *y-scale,* zoom in the coordinate grid around the point (a, b). In this case $a = 2$ and $b = 12$. Before you start zooming in, make sure that the x-coordinates of points *Me* and *Friend* are still as close to $x = 2$ as possible.

If you lose your axes, press the *Show Side Axis* button. If you lose point *Me* or point *Friend*, zoom out until you can see them again.

5. To use both sliders at once, use your **Arrow** tool to select the point at the end of each slider, then click on either point and drag. Familiarize yourself with what the sliders do and then zoom in until point *Me* or point *Friend* is on the edge of your sketch.

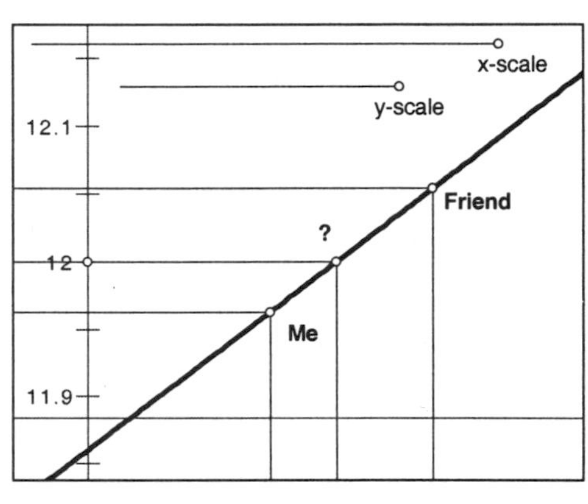

If $q(x)$ disappears, zoom out until it reappears. Select the plot and choose **Edit | Properties.** On the Plot panel, change the domain to a small interval around $x = 2$.

6. Drag point *Me* until you are as close as possible to $x = 2$, but keep your x-coordinate less than or equal to 2. Do the same with point *Friend,* but make sure its x-coordinate stays greater than or equal to 2.

Q5 How close to $x = 2$ can you get this time? What y-coordinate are you at now? What about your friend?

Q6 What y-value do you expect to see now at $x = 2$? (Does your left-hand expectation equal your right-hand expectation?)

Q7 Did zooming in strengthen or weaken your expectations?

You could keep zooming in and get closer and closer to $x = 2$ but your expectation, your limit, will stay the same. You then say that the limit of $q(x)$ or the limit of $\frac{x^3-8}{x-2}$ as x approaches 2 is 12, or, in shorthand, $\lim_{x \to 2} \frac{x^3-8}{x-2} = 12$. Notice that the reality, $q(2)$, is still undefined and you can't just substitute $x = 2$ into the function to get this limit.

Let's try another one. In the next few steps, you'll look for the limit of the function $\frac{\sin(x)}{x}$ as x approaches 0, or $\lim_{x \to 0} \frac{\sin(x)}{x}$. This function definitely has a problem at $x = 0$ and isn't defined there. But it might have a limit just like the above example, and then again, it might not.

7. Go to page 2 in your document. Here $q(x) = \dfrac{\sin(x)}{x}$.

Q8 Looking at the plot of $q(x)$ at $x = 0$, what y-value would you expect?

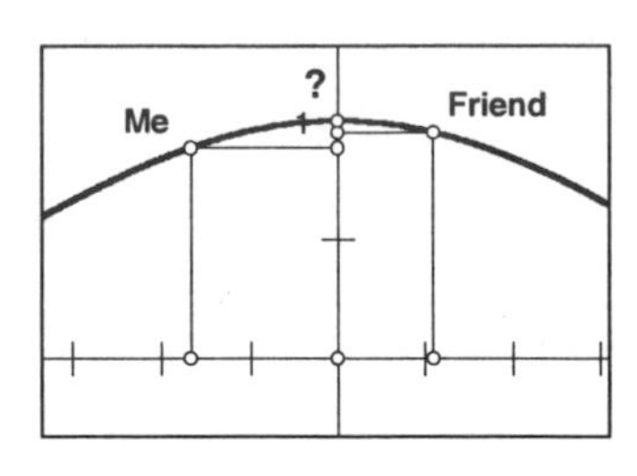

8. Drag point *Me* until you are as close as possible to $x = 0$, but keep to the left of 0. Do the same with point *Friend,* but make sure it stays to the right of 0.

If $q(x)$ disappears, zoom out until it reappears. Select the plot and choose **Edit | Properties.** On the Plot panel, change the domain to a small interval around $x = 0$.

9. Zoom in and see whether or not your initial guess in Q8 holds up.

Q9 Does your initial guess hold up? Does the left side agree with the right? What is the limit?

Now let's do a tricky one. In the next few steps, you'll look for the limit of the function $f(x) = 3x + \dfrac{0.01 \cdot |x-2|}{x-2}$ as x approaches 2, or $\lim\limits_{x \to 2}\left(3x + \dfrac{0.01 \cdot |x-2|}{x-2}\right)$. This function also definitely has a problem at $x = 2$ and isn't defined there. But it might have a limit just like the other examples—then again, it might not.

10. Go to page 3 in the document. Here is the plot of your new function.

11. Repeat the process that you used above. Make an initial guess, then drag point *Me* until you are as close as possible to $x = 2$, keeping to the left of 2. Do the same with point *Friend,* but make sure it stays to the right of 2.

Q10 Does your initial guess hold up? Does the left side agree with the right? What is your expectation at $x = 2$?

If you can't choose the **Properties** box, that means that something other than the function is selected. Click once in an empty area of your sketch, then select the function again.

12. Zoom in at $x = 2$ until your x-side axes go from about 1.97 to 2.04. Point *Me* and point *Friend* will look like they "fall off" the function if you take them close to $x = 2$. To take care of that, select the function plot and choose **Properties** from the Edit menu. Choose the Plot panel and change the domain for f to [1.92, 2.08].

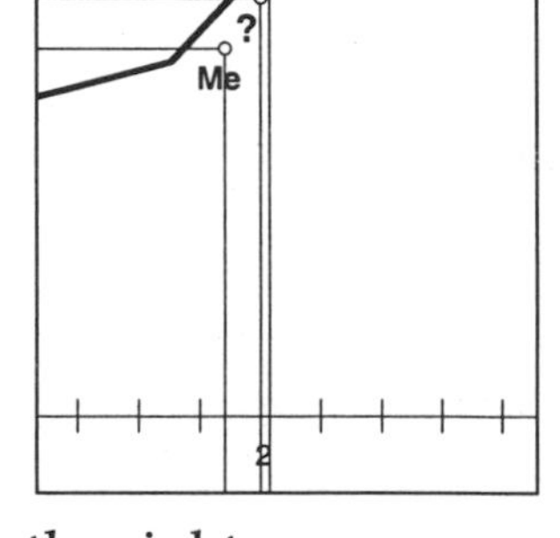

Q11 Try getting as close to $x = 2$ as possible again, using either point. Again make sure point *Me* stays on the left of $x = 2$ and point *Friend* stays on the right.

Q12 What is the biggest y-value you can get on the left-hand side before point *Me* jumps up? (This is the left-hand limit.) Can you ever get a y-value of 5.995?

Q13 What is the smallest y-value you can get on the right-hand side before point *Friend* jumps down? (This is the right-hand limit.) Can you ever get a y-value of 6.005?

Q14 Does the left-hand limit agree with the right-hand limit?

Did you get no for Q14? If you didn't, try this again. The function should jump from a y-value very close to 5.99 (but a little less than 5.99) to a y-value very close to 6.01 (but a little more than 6.01). This function does not have a limit at $x = 2$ because what you expect to see from the left (5.99) is *not* what you expect to see from the right (6.01). The left- and right-hand limits exist, but the limit itself does not. This is like having a canyon across your trail instead of a simple pothole. Potholes can be fixed so that the trail is smooth again. Canyons are a different problem altogether!

Explore More

Use page 4 to investigate *limits at infinity.* What y-value does the function approach if x becomes infinitely huge? Here $q(x)$ is back to the function $\frac{\sin(x)}{x}$. You are going to zoom out this time, so the sketch has been centered at (0, 0).

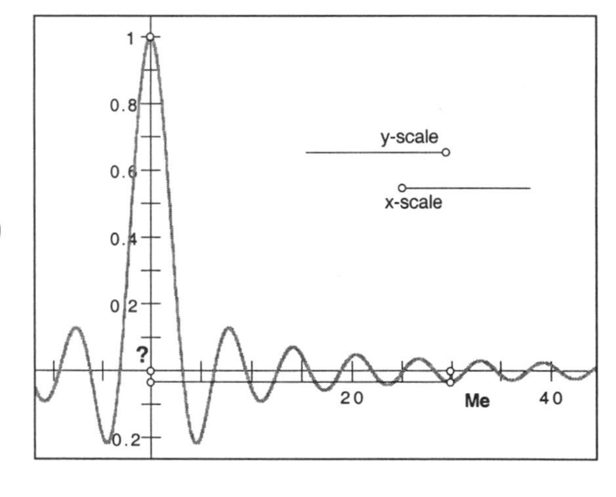

1. Zoom out until your window goes from –40 to 40 on the x-axis.
2. Select point *Me* and slide it along to about $x = 30$.

Q1 What is your approximate y-value at $x = 30$?

Q2 What is happening to the y-values as x gets bigger and bigger?

3. Zoom out until your window goes from –400 to 400 on the x-axis and move point *Me* to about $x = 300$.

You might need to zoom in on the y-scale to see what is happening around $x = 300$.

Q3 What is your approximate y-value at $x = 300$?

Q4 What is happening to the y-values as x gets bigger and bigger?

If you'd like to zoom out farther you'll need to change the domain of the plot. Select the plot, and then choose **Edit | Properties.** On the Plot panel, change the domain to some values farther out, such as –1000 to 1000.

4. Calculate the value of the function at $x = 1000$ by choosing **Calculate** from the Measure menu and entering $q(1000)$.

Q5 Based on your calculation and your plot, what do you think the limit is as x becomes infinitely big?

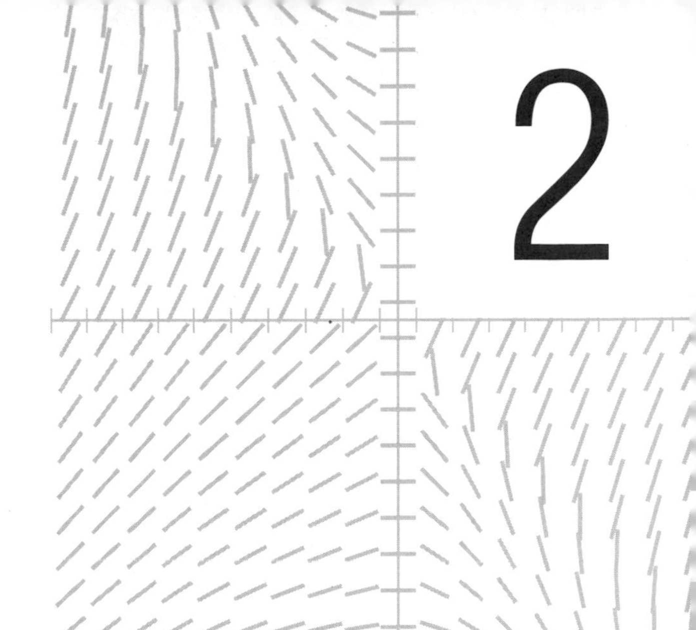

2

Exploring Limits

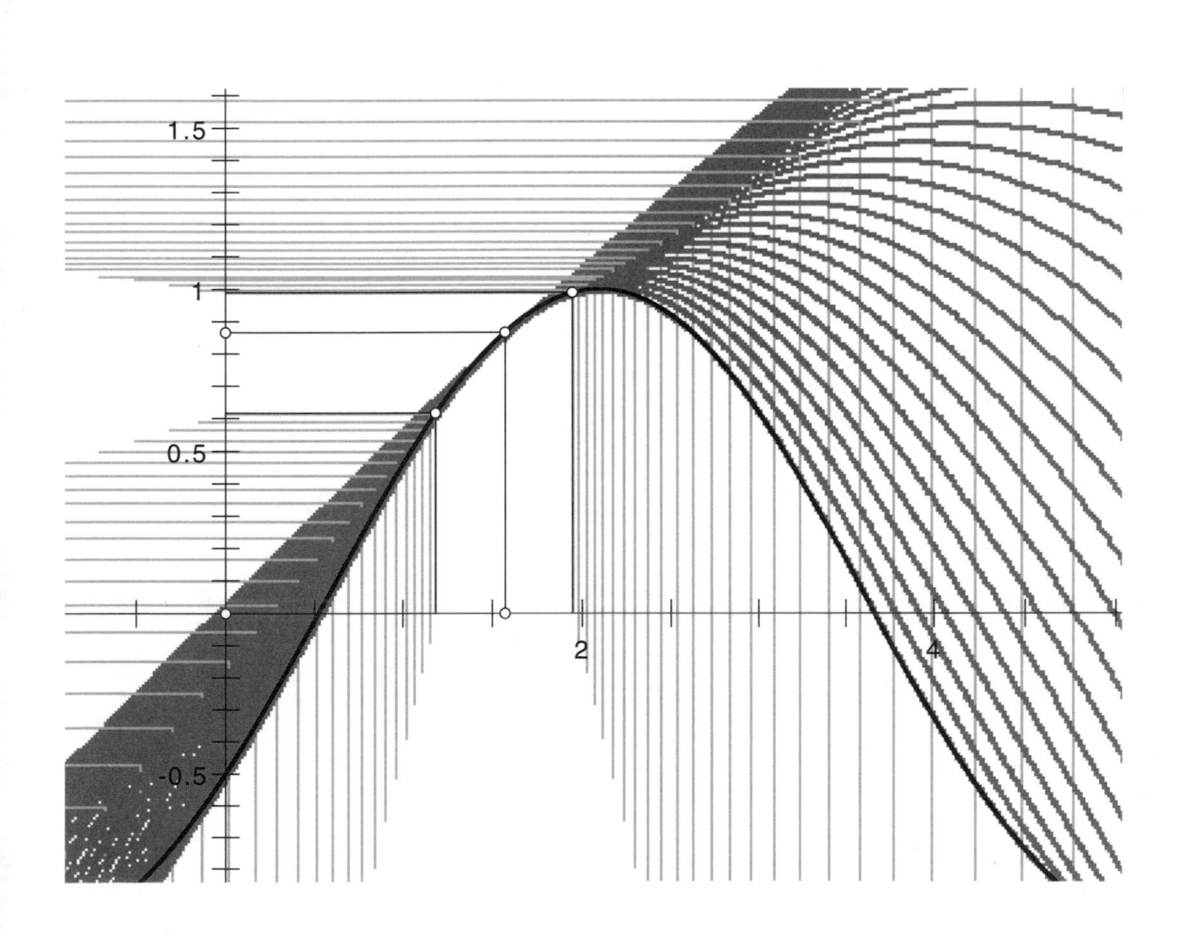

Continuous or Discontinuous? Name(s): ____________

As you've studied functions, you've learned about the different kinds of behavior functions can have. Some functions are constant, some are increasing, and some are decreasing. Some functions are linear and some are nonlinear. Some functions are defined on an unlimited domain, and others (such as the square root function) have a limited domain.

In this activity you'll look at a different aspect of the behavior of a function, an aspect which is particularly important in calculus: whether a function is continuous or discontinuous.

Sketch and Investigate

Imagine that you are a mountain climber making your way up Mt. Everest.

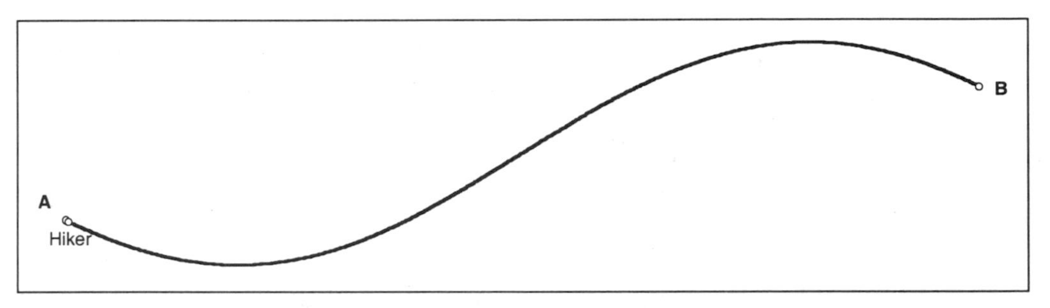

1. **Open** the sketch **Continuity.gsp** in the **Exploring Limits** folder. This sketch shows a rough profile of a glacier crossing, but without enough detail to see every little bump and dip.

This is not a particularly steep portion of your climb. Do you expect to have any difficulty hiking from point A to point B?

2. Press the *Start Hiking* button to begin the hike across the glacier. Observe what happens as the hiker heads for point B from point A. Press the *Show Details* and *Hide Rough Profile* buttons to show why you were rudely surprised.

Just as you must be vigilant about discontinuities in the surface of a glacier when you're hiking, you must also be vigilant about discontinuities in functions when you're doing calculus. You'll explore three different types of discontinuities in this activity.

Click on the page tabs in the bottom-left corner of the window to switch between pages.

3. Go to page 2 to explore the first type of discontinuity.

Point P was constructed on a discontinuous function. By slowly moving point P, you can trace out the function and find the discontinuity. If you drag point P more slowly, your trace will be more accurate.

4. Move point P to trace the function and try to locate the discontinuity.

Q1 At what x-value does the discontinuity occur?

5. Select point P, and then choose a different color from the Color submenu of the Display menu.

6. Press the *Trace Function* button to move point P automatically. Trace out one complete cycle, and then press the button again to stop the trace. Compare this trace with yours. Did you get the entire function, or did you miss part of it?

Pressing the Esc key once also stops the animation.

7. Press the Esc key several times to clear your traces.

Let's explore the function's behavior near the discontinuity in detail. It will be useful to do this in two parts, exploring the behavior both to the left and to the right of the discontinuity.

8. Press the *Show Left Branch* button. This will allow you to explore the behavior to the left of the discontinuity.

9. Two points appear: point A on the x-axis and the corresponding point $(x_A, f(x_A))$ on the function f—labeled point Q. Drag point A left and right. See how far to the right you can drag it.

Q2 What happens to point Q when you drag point A all the way to the right?

Click once in an empty part of the sketch to deselect an object.

10. Move point A back to the left of the discontinuity so that point Q is showing, then deselect point A. Select point Q, then measure its coordinates by choosing **Coordinates** from the Measure menu.

Q3 What happens to the coordinates of point Q when you drag point A as far right as it can go?

You can use animation to move point Q very close to the discontinuity—if you move point A very slowly.

11. To set this up: Select point A and choose **Animate Point** from the Display menu. Use the Motion Controller to decrease the speed to approximately 0.01. Then press the *Pause* button in the Motion Controller, and drag point A near the discontinuity.

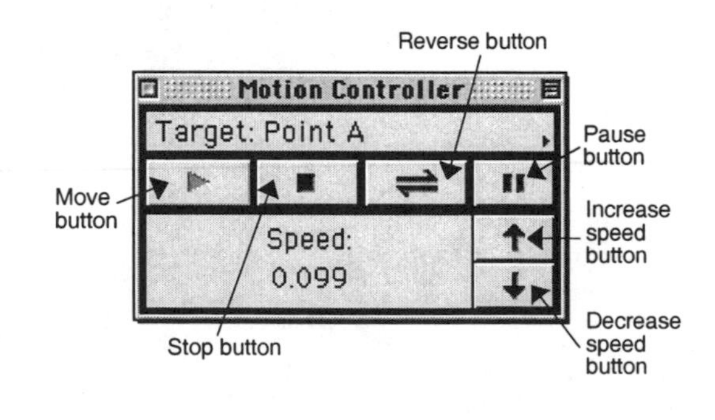

If point A is moving away from the discontinuity, press the *Reverse* button to change its direction.

12. Release the *Pause* button so that point A starts moving again, very slowly. Pause the motion again when point Q is very close to the discontinuity. By using the *Reverse* and *Pause* buttons and adjusting the speed you should be able to move point Q very close to the discontinuity.

Q4 Write down the coordinates of the closest point you achieved.

Q5 What do you think is the left-hand limit of the y-coordinate of point Q as the x-coordinate approaches 2?

Now let's explore the other side.

13. Hide the left branch and show the right branch. Two points appear: point B on the x-axis and the corresponding point $(x_B, f(x_B))$ on the function f—labeled point R. Measure point R's coordinates.

You'll create two parameters and a *Move* button to be more systematic about moving point R close to the discontinuity.

To enter a subscript type it in brackets. For End_x you would type End[x].

14. Choose **New Parameter** from the Graph menu. Name the first parameter End_x, give it a value of 2.1, and then click OK.

15. Make a second parameter with the name of End_y and a value of 0.

16. Select the measurements End_x and End_y, in that order. Then choose **Plot As (x, y)** from the Graph menu. Label this point *End.*

17. To create a button that will move point B to point *End,* with the **Arrow** tool, select point B and point *End,* in that order. Then choose **Edit | Action Buttons | Movement.** On the Move panel, choose **medium** speed and click OK.

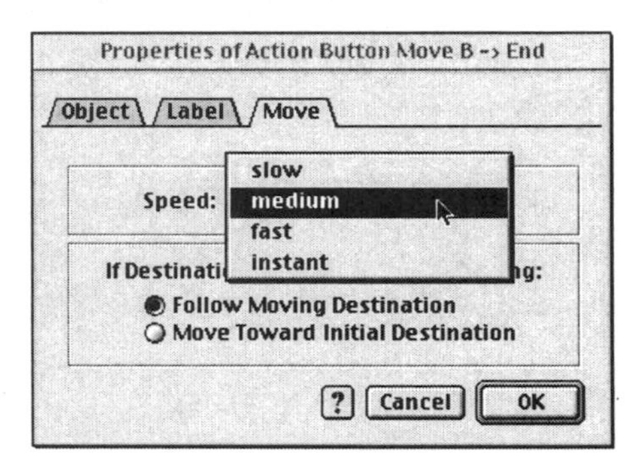

Make sure the Motion Controller is not on pause, otherwise nothing can move.

18. Click once in an empty spot to deselect the *Move* button. Then select point *End* and choose **Display | Hide Plotted Point** to hide it. Then press the *Move B→End* button.

Q6 What are the coordinates of point R?

19. Double-click on the measurement End_x. Change its value to 2.01. Then press the *Move B→End* button.

Q7 What are the coordinates of point R?

Repeat step 19, putting point *End* even closer, at (2.001, 0).

Q8 What are the coordinates of point R this time?

Q9 What is the right-hand limit of the y-coordinate of point R as the x-coordinate approaches 2? Is this the value that you found for the left branch? Does the limit exist at $x = 2$?

20. Go to page 3 to see a different type of discontinuity. Trace this function by dragging point P_x (*not* point P) or by pressing the *Trace Function* button. Do you see a discontinuity as you trace?

21. Clear your traces and show the function plot. Can you see a discontinuity now?

There is a discontinuity here, but it's even narrower than the crevasse that the hiker fell into.

22. Press the *Move* $P_x \rightarrow 2$ button. Can you see a discontinuity now?

Q10 What happens to point P when point P_x is at exactly $x = 2$?

23. Hide the function plot and erase traces. Then show the left branch, measure the x- and y-coordinates of point Q, and determine the limit of the y-value as the x-value approaches 2 from the left.

24. Hide the left branch, show the right branch, measure the coordinates of point R, and determine the limit of the y-value as the x-value approaches 2 from the right.

Q11 What is the limit of the y-value as x approaches 2 from the left? From the right? What is the limit as x approaches 2?

25. Go to page 4 to see the last type of discontinuity. Trace this function by dragging point P or by selecting point P and choosing **Animate** from the Display menu. Find the discontinuity as you trace.

26. Show the function plot and use point A to find three x- and y-values just to the left of the discontinuity.

Q12 List the three closest points you found in step 26 in numerical order.

27. Find three x- and y-values just to the right of the discontinuity.

Q13 List the three closest points you found in step 27 in numerical order.

Q14 What is the limit of the y-value as x approaches –1 from the left? From the right?

On pages 2, 3, and 4 you explored three different types of discontinuities in functions. Answer the following questions about these discontinuities. Use a separate sheet of paper, and write in complete sentences.

Q15 Give a name to each type of discontinuity. First come up with a descriptive name of your own, and then do some research to find out what names other people have used.

Q16 What are the similar features of all three types of discontinuities?

Q17 For each type of discontinuity, describe the features that make it different from the other two.

Q18 Which type of mathematical discontinuity best describes the situation of the hiker who fell into the crevasse? Justify your answer!

Explore More

Use Sketchpad to graph each function listed below. For each one, locate and explore its discontinuity and decide which type of discontinuity it is. (Some functions may have more than one discontinuity.)

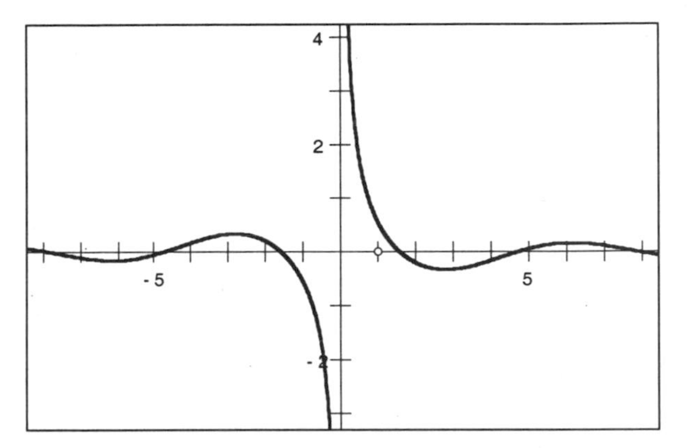

1. $f(x) = \text{trunc}(x)$

2. $f(x) = \left|\dfrac{1}{x-1}\right|$

3. $f(x) = \dfrac{x^2 - 3x + 2}{x - 2}$

4. $f(x) = \dfrac{\sin(x)}{x}$

5. $f(x) = \dfrac{\cos(x)}{x}$

6. $f(x) = \tan(x)$

7. $f(x) = \dfrac{x^2 - 3x + 2}{|x - 2|}$

Delta, Epsilon, and Limits **Name(s):** ____________________

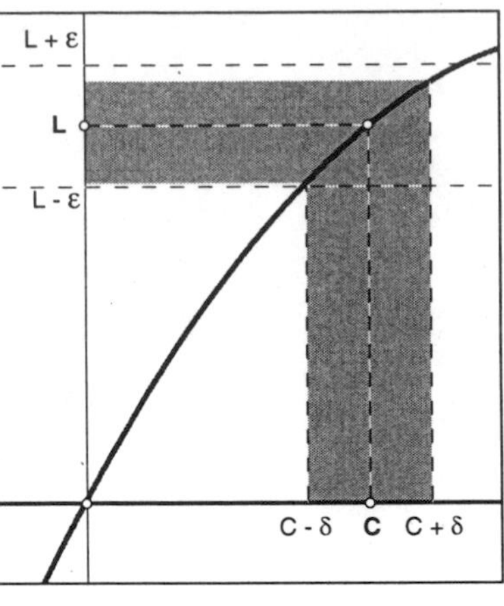

You know that a function $f(x)$ has a limit, L, at $x = C$ if you can make the y-coordinate of the function as close as you want to L by choosing an x-coordinate really close to $x = C$.

More formally, L is the limit of $f(x)$ as x approaches C if and only if, for any positive number epsilon, ε, there is a positive number delta, δ, such that if x is within δ units of C ($x \neq C$), the y-value of f is within ε units of L.

In this activity, you'll explore this definition visually. For a given function and for given values of C, L, and ε, you'll adjust the value of δ to test the definition.

Sketch and Investigate

The constants a, b, and c are controlled by sliders. You can change the function by changing the values of the constants a, b, and c.

1. **Open** the sketch **EpsilonDelta.gsp** in the **Exploring Limits** folder. This sketch shows the graph of $f(x) = ax^2 + bx + c$.
2. Experiment with the *epsilon* slider, which determines an interval on the y-axis about L. This interval is represented by a vertical red segment containing all y-values within ε of L.
3. Experiment with the *delta* slider, which determines an interval on the x-axis about C. This interval is represented by the width of a vertical blue polygon and contains all x-values within δ of C.

The intersection of the left and right edges of this blue polygon with the function determines another polygon (horizontal this time). The height of this second green polygon represents the y-values of the function when the x-value is restricted to within δ of C.

You can reset these constraints by pressing the *Show Parameters* button. Double-click on a, b, or c to change its value.

4. Set the sliders for a, b, and c to $a = 0.6$, $b = 2.5$, and $c = -2$ by pressing the *Set Quadratic* button.
5. Press the *Case 1* button to set the values of C, L, and ε.

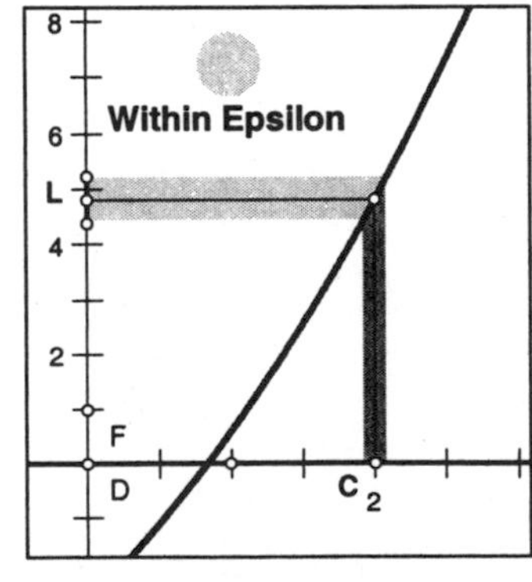

The definition of a limit will be satisfied graphically if you can adjust δ—using the *delta* slider—in such a way that all the possible y-values represented by the horizontal green polygon fall within this red interval (that is, they fall within ε of L or in the interval $[L - \varepsilon, L + \varepsilon]$).

Use keyboard arrows to move one pixel at a time.

6. Adjust the *delta* slider until the horizontal green polygon falls *just* within the red $[L - \varepsilon, L + \varepsilon]$ interval on the y-axis. (The circular indicator changes from red to green when *all* possible y-values lie within the interval.)

Q1 What is the largest value you found for δ for this value of ε? Record the values of C, L, ε, and δ in a table on a separate sheet of paper.

7. Move the *epsilon* slider to set ε to roughly half its previous value. Adjust δ to work with this new value of ε. Record this new set of results in your table.

Q2 When you set ε to half its previous value, did your value for δ reduce by half as well? Try other fractions of ε, such as 2/3 or 2/5.

Drag the unit points to resize your window and then press the *Case* button again to reset the value of ε.

8. Press the *Case* 2 button, and repeat steps 6–7 and Q1–Q2. Add your results to the table you made in Q1.

9. Repeat for Cases 3 and 4, adding your results to the table each time.

Q3 Was it possible in each case to set a value of δ that satisfied the definition of a limit? If not, note the values for C, L, and ε for which it was not possible. Did your effort fail because L does not exist, or because you could not adjust δ correctly? What can you conclude about the function as x approaches C in each case?

Q4 Do you think it would be possible to find such a value of δ graphically for every possible value of C and ε? If not, what values of C and ε would not work?

You will need to resize your window for some of these cases. Don't forget to press the *Case* button after resizing to reset the value for ε.

10. Go to page 2 of the document and follow the same steps described above. For each case, adjust the *delta* slider so that the resulting y-values fall within ε of the limit L, and record your results in a table. Record just one set of results for each case.

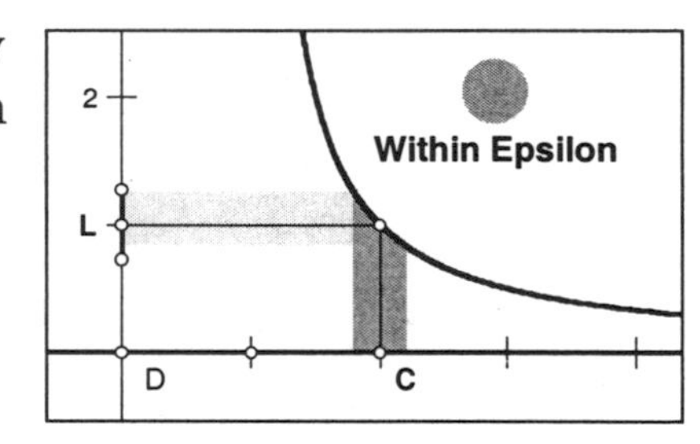

Q5 Was it possible in each case to set a value of δ that satisfies the definition of a limit? If not, note the values for C, L, and ε for which it was not possible. Did your effort fail because L does not exist, or because you could not adjust δ correctly? What can you conclude about the function as x approaches C?

Explore More

To graph a new function, double-click on the expression for $f(x)$, delete the current expression, and enter your own.

Go to page 3 of the document and experiment with the fourth-degree polynomial there or graph a new function.

Go to page 4 of the document and experiment with the trigonometric function there or graph a new function.

How Close Do You Go?

Name(s): ____________________

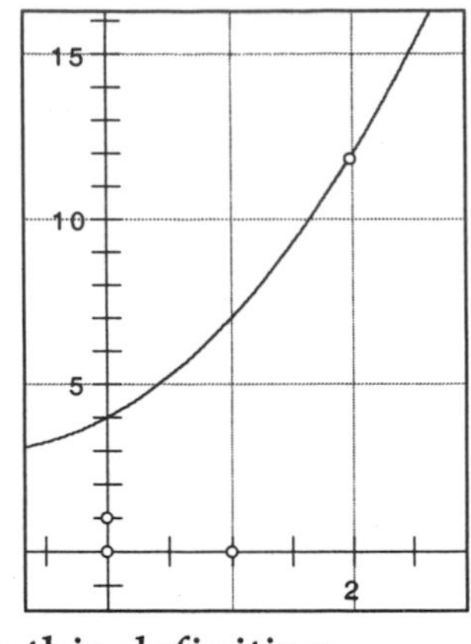

You have seen that the y-value of a function can approach a limiting value as x approaches a constant c even if f is undefined. For example, if you plot the function $q(x) = \dfrac{x^3 - 8}{x - 2}$, you get a parabola with a gap or hole at $x = 2$. But as x approaches 2, the y-value of the function approaches 12.

You have learned that L is the limit of a function f as x approaches c if you can make the y-value of the function as close as you want to L by making x sufficiently close to c. In this activity, you will explore this definition graphically for various functions.

Sketch and Investigate

1. **Open** the document **Limits2.gsp** in the **Exploring Limits** folder.

On page 1, you have the plot of the function $f(x) = \dfrac{x^2 - 9}{3x - 9}$, which has a hole at $x = 3$. Points P and N will allow you to evaluate the function for x-values near $c = 3$.

Q1 Try to find $f(3)$ by direct substitution. What happens?

Q2 What limiting value does the plot suggest as x approaches 3? Check your answer by simplifying the expression for f and substituting $x = 3$. Do you get the same value for the limit?

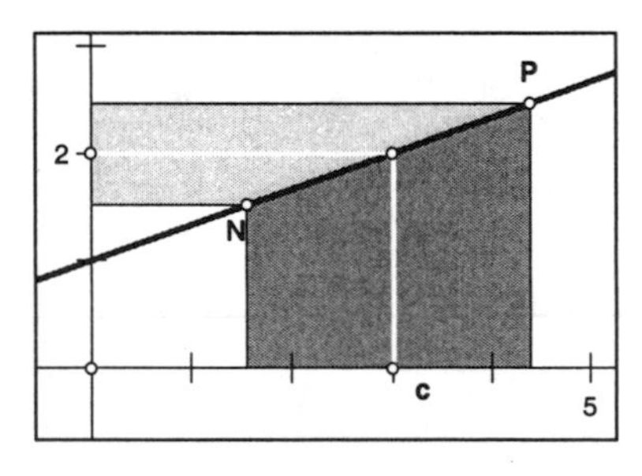

In order to show graphically that the definition holds, you need to show that you can make $f(x)$ as close as you wish to $L = 2$ by making x sufficiently close to $c = 3$.

Q3 Suppose someone tells you to keep $f(x)$ within 0.5 units of the limit 2. How close must x be to $c = 3$ to make the y-values of the function fall within that interval?

Check that $a = 3$ and $b = 2$. If they don't, edit them by double-clicking on their measurements.

To get a little closer you can zoom in around this point even though $f(x)$ is not defined at $x = 3$. The x- and y-*scale* sliders zoom the coordinate grid around the point (a, b) for the values of a and b defined in the sketch.

2. Drag points P and N as close as you can to the limit.

If you want to use both sliders at once, select the endpoint of each slider, then select either point and drag.

3. Zoom in using the sliders until you lose your axes. Then press the *Show Zoom Axes* button, and then the *Hide Main Axes* button. Zoom in again until the x-axis goes from about 2.2 to 3.8 and the y-axis from about 1.5 to 2.5.

This will be a rough estimate because you are using the tick marks on the zoom axes and eyeballing it.

4. Adjust the location of points P and N to find out how close you have to make x to 3 so that $f(x)$ is within 0.1 of the limit 2.

5. Measure the coordinates of points P and N to make sure that all of the horizontal rectangle is within 0.1 of the limit.

Q4 How close do you have to make x to 3 so that $f(x)$ is within 0.1 of the limit 2?

Drag points P and N as close as you can to the limit before you zoom in.

6. Adjust the sliders and the location of points P and N to show how close you must make x to 3 so that $f(x)$ is within 0.05 of 2.

Q5 How close do you have to make x to 3 so that $f(x)$ is within 0.05 of 2?

You can calculate $f(x_P) - L$ and $L - f(x_N)$ to check these limits as well.

Q6 How close do you have to make x to 3 so that $f(x)$ is within 0.01 of 2? Within 0.001 of 2?

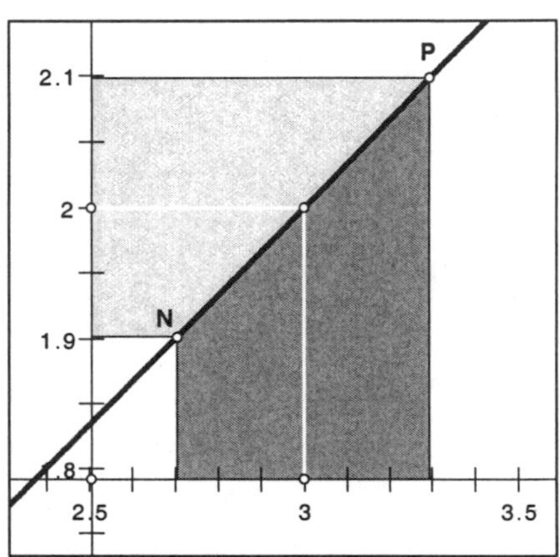

As you saw above, to demonstrate graphically that a limit exists, you show that you can make x close enough to c so that the value of $f(x)$ will always lie within a narrow band around the limit. In the last activity you learned that this distance from L is called *epsilon*, or ε, and the distance from c is called *delta*, or δ. In the example shown here, $\varepsilon \approx 0.1$ and $\delta \approx 0.3$.

Q7 Look back at your answers to the previous questions. Does there appear to be a relationship between the value of ε and the value of δ? If so, what is this relationship?

Q8 If you look at the sketch above, the interval around $c = 3$ is symmetric going from about $3 - 0.3$ to about $3 + 0.3$. Why do you think this was true for all of your intervals above?

Let's try another function to see if we usually get a symmetric interval around our target value $x = c$.

7. Go to page 2 of the document. Here $f(x) = 2 - 4^{x-1}$ and the limit as x approaches 0.5 is 1.5. Zoom in and adjust point P and point N so that the y-values are within 0.1 of the limit.

Q9 How close do you need to make x to 0.5 on the left side of point C? On the right side of point C? What would your δ be in this case?

Now we have the hypothesis that for exponential functions (without a discontinuity) the intervals won't be symmetric. But this is just one example of a function—let's look at one more polynomial.

With the Calculator open, click on the expression for $f(x)$ and then on measurement c.

8. Go to page 3 of the document. Choose **Calculate** from the Measure menu and calculate $f(3)$.

9. Choose the measurements for c and $f(c)$ in that order and choose **Plot As (x, y)** from the Graph menu.

Q10 Is the function f continuous at $x = 3$?

Q11 What value does the plot suggest for the limit as x approaches 3?

10. On this page, two sliders control the location of points P and N. The slider labeled h_N controls the distance from point N to point C on the left and the slider labeled h_P controls the distance from point P to point C on the right. Experiment with the sliders to see how point P and point N move.

Once you've zoomed in, press the *Show Zoom Axes* button and then press the *Hide Main Axes* button.

11. Zoom in and adjust each slider to see how close you need to make x to 3 in order to make $f(x)$ within 0.1 of the limit. Then adjust the sliders to see how close you must make x to 3 in order to make $f(x)$ within 0.05, 0.01, and then 0.001 of the limit. Make a table of your δ and ε values.

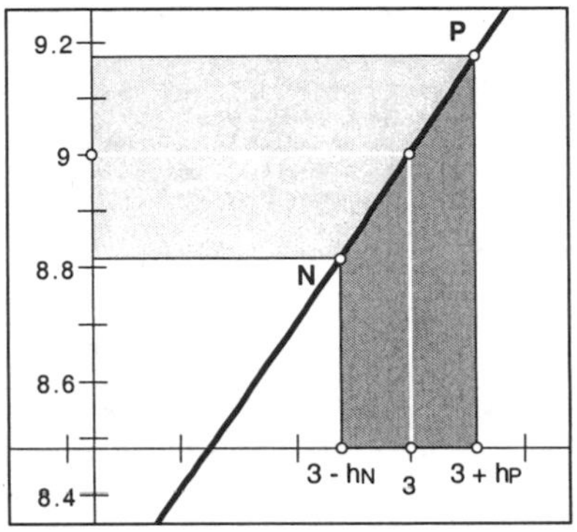

Q12 Were your intervals around $c = 3$ symmetric? Can you find a relationship between the values that you found?

This last example had intervals that were not symmetric. (Go back and look again if you answered yes to Q12.) It turns out that most of the time, they aren't. So how was the first function different from the other two?

Let's look at one more example.

12. Go to page 4 of the document. Choose **Calculate** from the Measure menu and calculate $f(3)$.

13. Choose the measurements for c and $f(c)$ in that order, and choose **Plot As (x, y)** from the Graph menu.

Q13 Is the function f continuous at $x = 3$? What value, if any, does the plot suggest for the limit as x approaches 3?

14. Try as before to adjust each slider to see how close you need to make x to 3 in order to make $f(x)$ within 0.1 of your prediction, or, try to show why there is no limit using 0.1 as your ε.

Q14 What happened when you tried to do step 14, and how is this function different from the previous examples?

Q15 What are the left- and right-hand limits as x approaches 3? Explain why no single number can be the limit of $f(x)$ as x approaches 3.

The four examples above covered each of the classifications below except one. Figure out which classification has not been covered yet, and match the four functions above with their classifications.

A. Defined at $x = c$, limit exists at $x = c$, continuous at $x = c$.

B. Defined at $x = c$, limit exists at $x = c$, not continuous at $x = c$.

C. Defined at $x = c$, limit does not exist at $x = c$, not continuous at $x = c$.

D. Undefined at $x = c$, limit exists at $x = c$, not continuous at $x = c$.

E. Undefined at $x = c$, limit does not exist at $x = c$, not continuous at $x = c$.

Explore More

Here is a function that satisfies the missing category above:

$$f(x) = \frac{|x-3|}{x-3} + x$$

1. Go to page 5 and confirm that this function satisfies the conditions—it is undefined at $x = 3$ and the limit does not exist for $x = 3$.
2. Double-click on measurement c and change it to $c = 0$. Change b to 1.
3. Double-click on the expression for $f(x)$ and change it to $f(x) = 1 + x\cos(1/x)$. Investigate the behavior of $f(x)$ as x approaches 0, and then compare it to the function $f(x) = 1 + \cos(1/x)$.
4. Your function may get choppy as you zoom in. To fix that, select the plot of f, then choose the Plot panel from **Properties** in the Edit menu, and change the domain to a smaller interval around 0.

Q1 Does the limit as x approaches 0 exist for either of these functions?

Q2 How are these functions different in this region? How are they the same?

Q3 As x becomes infinitely large, does the y-value seem to approach a limiting value for either of these functions? If so, what value? If not, describe why not.

Q4 How could you use Sketchpad to demonstrate graphically that a limit exists or doesn't exist as x becomes infinitely large?

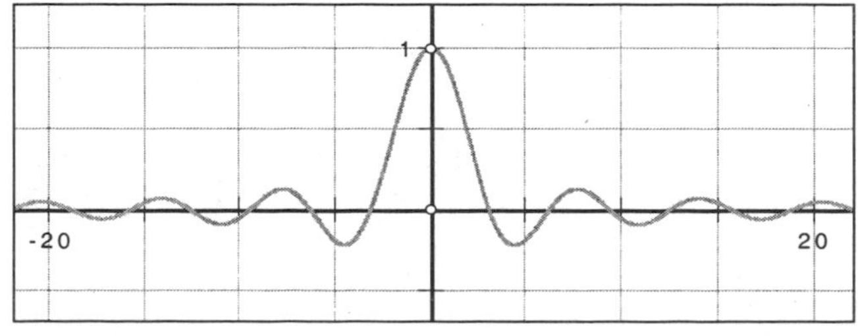

Slope and Limits

Name(s): ______________________

Limits are a powerful tool for describing the behavior of functions at a particular point or as x becomes infinitely large. Limits also provide the missing link we need to answer this question: What is the instantaneous rate of change of the function at a point, or what is the function's slope at any given point?

In this activity you will explore how the limit can be used to move from average rates and secant slopes to the instantaneous rate and slope at a point.

Sketch and Investigate

1. **Open** the document **SlopeandLimit.gsp** in the **Exploring Limits** folder. On the first page you will find a linear function f with its plot and points P and Q. Point P is fixed by the parameter x_P. Point Q is free to move along the function plot.

Choose **Calculate** from the Measure menu and click on each measurement to enter it into the calculator. Don't forget parentheses.

2. Find the average rate of change between points P and Q by calculating the slope of line PQ:

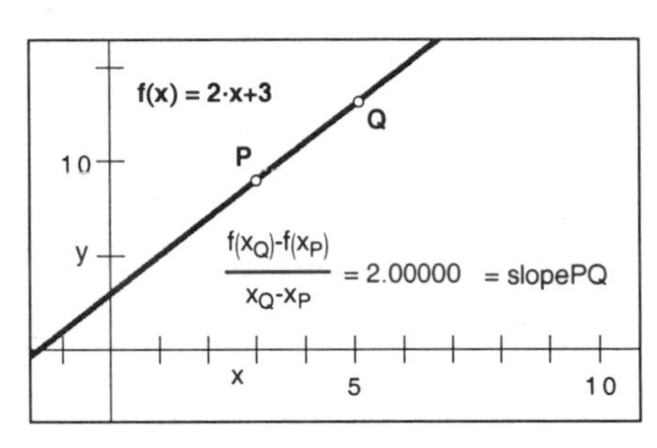

$$slopePQ = \frac{f(x_Q) - f(x_P)}{x_Q - x_P}$$

Label this measurement *slopePQ*.

3. Does your result make sense? Move point Q along the function plot, and confirm that the value of the calculation does not change.

The coordinates of point P are (3, 9), and because point Q is free to move along the function plot, its coordinates are $(x_Q, 2x_Q + 3)$. So the slope of the line from a fixed point P to point Q is *always:*

$$slopePQ = \frac{f(x_Q) - f(x_P)}{x_Q - x_P} = \frac{(2x_Q + 3) - 9}{x_Q - 3} = \frac{2x_Q - 6}{x_Q - 3} = \frac{2(x_Q - 3)}{x_Q - 3} = 2$$

The fact that the slope of a linear function at a point is indeed its usual slope should not have been surprising. But what is the slope at a point if the function is not linear? Is it even possible to ask such a question?

Double-click on the expression for $f(x)$ to edit the function.

4. Move point Q to the left of point P, and change $f(x)$ to x^2.

Select both points, and then choose **Line** from the Construct menu.

5. Construct a line between points P and Q, and give it a different color than the function plot by choosing **Display | Color.**

Q1 Move point Q along the function plot. Does the value of the measurement *slopePQ* stay constant as it did in step 3? If so, why do you think it stays constant? If not, describe its values as you move point Q from left to right.

Q2 What are the coordinates of point P? Point Q is not fixed, so its coordinates are not constants—what are they? Write an expression to calculate the slope of the line from point P to point Q. (See the example after step 3 if you get stuck.)

Your expression in Q2 should have been a simplified version of $\frac{f(x_Q) - f(3)}{x_Q - 3}$, or the average rate of change from point P ($x = 3$) to point Q.

This expression is also a function, and it can be examined using limits.

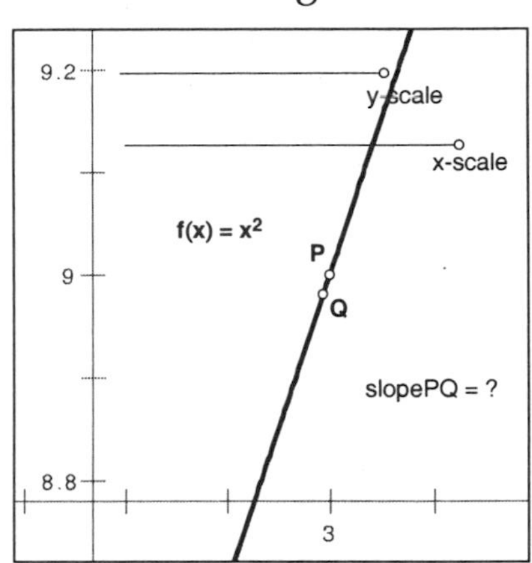

Q3 Find $\lim_{x \to 3} \frac{f(x_Q) - f(3)}{x_Q - 3}$ algebraically.

Q4 Move point Q as close as you can to point P. Pay close attention to the measurement *slopePQ*, and look for its limiting value. As Q approaches P, does the limiting value of *slopePQ* appear to graphically support your answer to Q3?

> In this document, the coordinate grid will dilate around the point $(x_P, f(x_P))$. To change the location of point P, edit the parameter x_P.

6. Zoom in by moving both the x- and *y-scale* sliders to the right. Point Q will appear to move away from point P, but note that the coordinates of point Q do not change.

7. Move point Q closer to point P.

Q5 What is the slope of the line PQ when Q is very close to P? Is this value approximately equal to the limit you calculated in Q3?

In the steps above, you examined, not the limit of the function f at a point, but the limit of a new function based on f. This new function is called the *difference quotient* at $x = 3$, but it is still the average rate of change of $f(x)$ at $x = 3$, or the slope of a secant line through point P.

We define *the slope of a curve at a point P* to be the limit of the slopes of these secant lines, PQ, as point Q approaches point P. This limit is also referred to as *the instantaneous rate of change of the function at the point P.* The limiting line itself can be used to visualize the limit.

Let's see why.

> If you can zoom in at a point until the function looks linear, we say that the function is *locally linear* at that point.

8. Zoom in until the function looks like a line and you can't see any difference between the function $f(x)$ and the line PQ. Move point Q as close as possible to point P.

9. Choose **SlopeTool** from **Custom** tools. This tool calculates the slope between *any* two points, so use it to make two new points anywhere on the "line." (Make sure the status message is "to Point on Function Plot.")

Q6 What value did the **SlopeTool** give you for the slope of the zoomed-in plot of the function? Select either new point and drag it anywhere in the window. Does this slope value change?

Q7 Explain in your own words why this limiting line can be used to visualize the limit.

Not only are limits useful for analyzing a function's behavior, you've also seen that they give us the missing links to define an instantaneous rate of change at a point, or a function's slope at a point. Defining an instantaneous rate of change is just one reason for building a difference quotient and finding its limit, as you will see in the next activity, "Area and Limits."

Exploration 1

There is another reason why we'd like to look at a function's local linearity at a point—working with a line is usually easier than working with a general nonlinear function. If we use the line instead of the function, how much error is there? To find out, we need to look at another point close to point P.

If you have moved point Q or have changed your window since step 8, repeat step 8 above. Then you can do this step.

1. Leaving point Q as close as possible to point P, zoom back out using both sliders until the x-axis goes from about –2 to 7 and the y-axis goes from about –2 to 12.

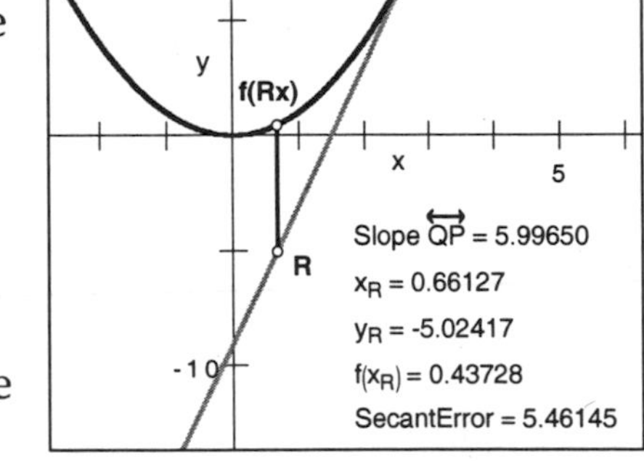

2. Using the **Point** tool, make a new point on the line PQ and label it R.
3. Measure the x-coordinate of point R by selecting **Abscissa(x)** from the Measure menu. Then measure its y-coordinate (**Ordinate(y)**).
4. Calculate $f(x_R)$ and plot the point $(x_R, f(x_R))$.
5. Construct the line segment between the point plotted in step 4 and point R.

Q1 Explain in your own words why this last segment graphically represents the error created if you used the y-coordinate of point R to approximate the function's y-value, $f(x_R)$.

Q2 Drag point R so that $x_R \approx 2$. What would the error be if you used the y-coordinate of point R to approximate $f(2)$?

6. Write a formula you can use to calculate this error. Create this calculation in your sketch and label it *SecantError*.

Q3 Move point R along the line PQ and determine how close you need to get to $x = 3$ to make the error less than 0.1, 0.05, and 0.001.

Q4 You have the point P and you calculated the limiting slope for this line above. Write the equation of this limiting line in point-slope form.

7. Check your answer by choosing **Plot New Function** from the Graph menu and entering the equation for your line from Q4. Zoom in and out to see if your line matches line PQ.

Q5 By hand (no calculators!), find the y-value for $x = 3.04$ using your line and then using your function. The calculation with the linear equation should have been much easier—what was the error?

Exploration 2

Go to page 2 to see an example of a function with points where f can't be locally linearized. Let's see what goes wrong at $x = 0$.

1. Choose the **Slope+Line** tool from **Custom** tools. This tool calculates the slope between two points and creates the secant line. Match point P (the origin) and make a new point with this tool on the function where $x \approx -4$.

2. Use this tool to make three more secant lines from point P to $f(x)$ where $x \approx -3, -2$, and -1.

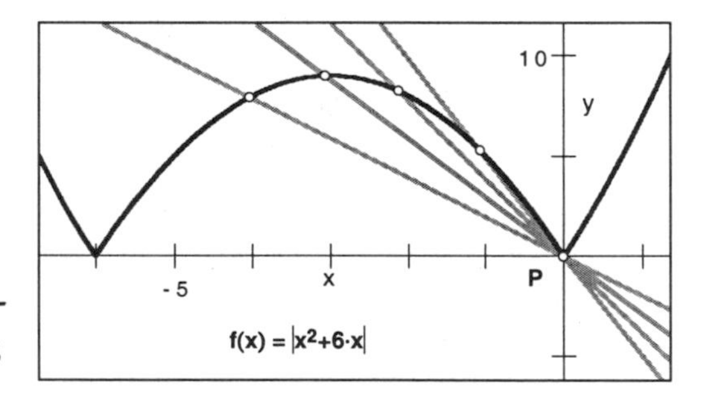

Q1 Using the four slope measurements calculated by the **Slope+Line** tool, estimate the left-hand limit for the secant slopes as x approaches 0.

3. Select each of the secant lines and choose **Display | Hide Lines.**

4. Use the **Slope+Line** tool to make secant lines from point P to $f(x)$ where $x \approx 0.5$, 1, 1.5, and 2.

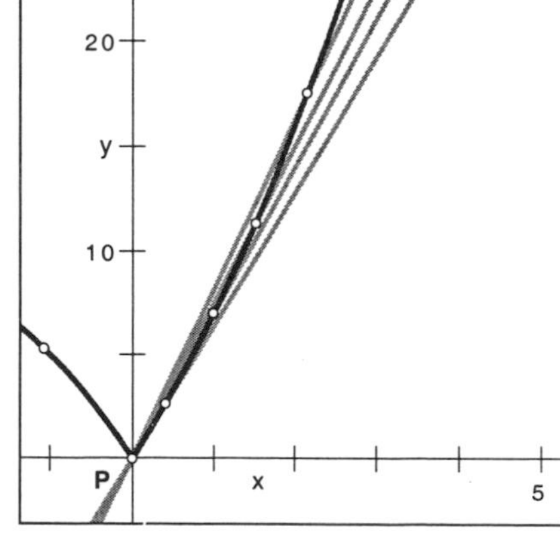

Q2 Using the four slope measurements calculated by the tool, estimate the right-hand limit for the secant slopes as x approaches 0.

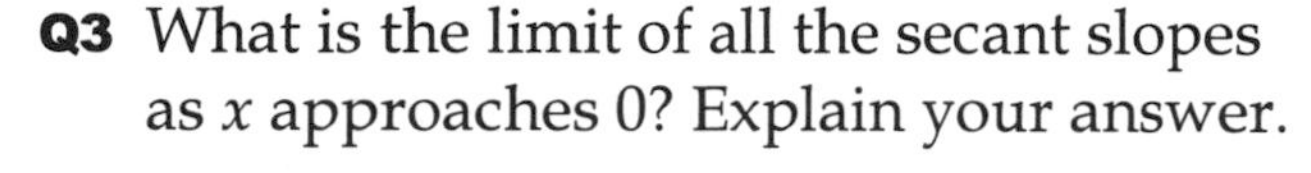

Q3 What is the limit of all the secant slopes as x approaches 0? Explain your answer.

5. Pick any of the new points created by the **Slope+Line** tool, and drag it across point P. Describe what the secant line does as the point moves through P.

Q4 Explain in your own words why f is not locally linear at x.

Exploration 3

1. Go to page 3 in the document, and change the function to $f(x) = \sin(x)$, and x_P to π. (Try $x = \pi/2$ as well.)

Q1 What limit do you need to evaluate to determine the instantaneous rate of change of sin(x) at $x = \pi$? Zoom in to estimate the limit.

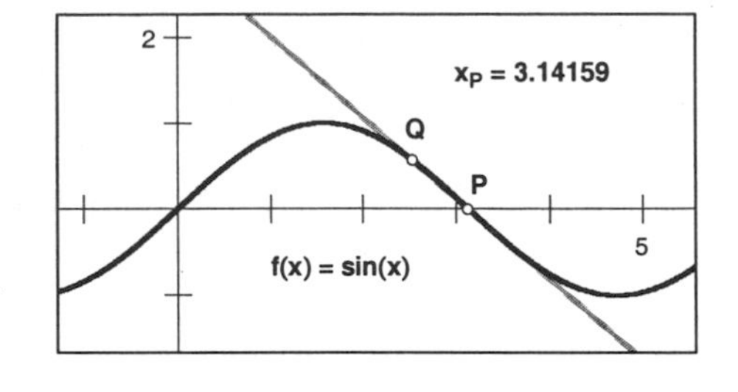

2. Zoom back out. Change the function to $f(x) = \ln(x)$ and x_P to 4. (Try $x = 3$ and 5 as well.)

Q2 What limit do you need to evaluate to determine the instantaneous rate of change of ln(x) at $x = 4$? Zoom in to estimate the limit.

3. Zoom back out. Change the function to $f(x) = \sqrt{x}$ and x_P to 4. (Try $x = 2$ and 5 as well.)

Q3 What limit do you need to evaluate to determine the instantaneous rate of change of $\sqrt{x}$ at $x = 4$? Zoom in to estimate the limit.

Q4 You may be familiar with situations where limits do not exist. Can you think of a function with a point where the instantaneous rate of change, as defined above, will not exist? Write the function and the accompanying limit, and show that this limit does not exist. (*Hint:* Think about the slope of the line you used to visualize the limit.)

Area and Limits

Name(s): ____________________

The area of the shaded region on the right looks to be a little less than 0.5 square units. But how much less? If you had to come up with one number for the most accurate measurement of the area, what would it be? How would you find that number? How might you convince someone that your approximation is the most accurate? This activity will help you begin to answer these questions.

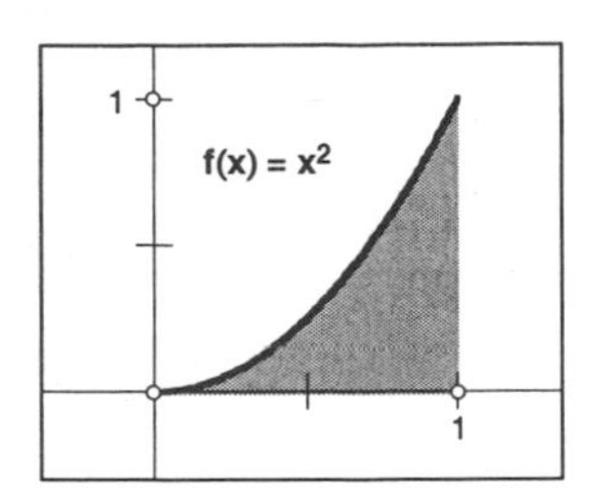

Sketch and Investigate

1. **Open** the document **AreaLimits.gsp** in the **Exploring Limits** folder. On page 1 there is a constant function $f(x) = 4.9$ on the domain $x = 0$ to $x = 10$.

Q1 Using geometry, find the area of the shaded region under the curve $f(x) = 4.9$ from $x = 0$ to $x = 10$.

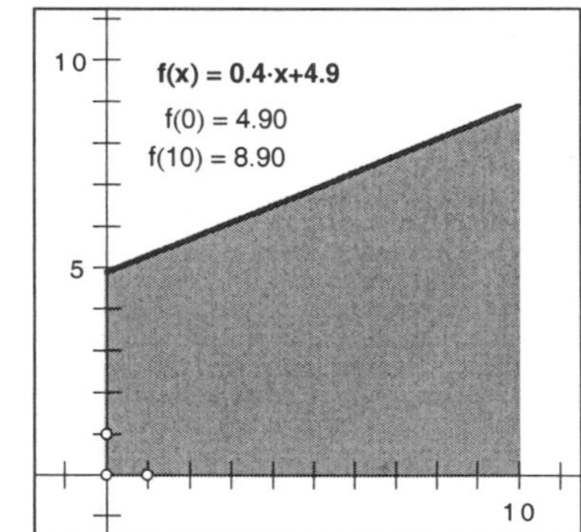

2. Go to page 2 of the document. On this page, $f(x) = 4.9$ at $x = 0$, but this time, f increases at a constant rate on its domain [0, 10] to a value of 8.9 at $x = 10$.

Q2 Calculate the area of this shaded region by using the area formula for a trapezoid.

Unfortunately it's not always possible to find the number of square units of area by using a formula from geometry.

3. Press the *Next Page* button to go to page 3. On this page, you will find the plot of the function $f(x) = 0.1(x - 3)^2 + 4$.

Q3 This function has the same y-values as the previous linear function at $x = 0$ and $x = 10$. How do you think the area of the shaded region will compare to the area of the region on the previous page?

If your shaded region moves outside of the window, drag an axis or the origin until you can see the whole region.

4. Drag the unit point at (1, 0) to the right until there is a grid line at every integer. Use the grid to count the "boxes" (squares of area 1) and estimate the area of the region.

Q4 What is your estimate for the area of the shaded region if you use only whole boxes? Is this an overestimate or an underestimate?

Q5 What could you do to make a better or more accurate estimate?

5. Press the *Next Page* button to see a really large overestimate—one rectangle with a base of 10 units and a height of 8.9 units.

Here we have one rectangle on the interval [0, 10] and n defined as the number of rectangles, so here $n = 1$. The right endpoint of the interval [0, 10] is used to find the height, which is the y-coordinate at $x = 10$, or $f(10)$, so this rectangle is called a *right rectangle*. If we use the left endpoint of the interval ($x = 0$) to find the height, or $f(0)$, the rectangle is called a *left rectangle*.

Point n on the left of the bar will make left rectangles.

6. Drag the slider for n to the left, past the bar, until $n = 1$ to show one rectangle using the left endpoint of the interval. This gives an estimate that is much too small—a rectangle with a base of 10 units, but this time with a height of 4.9 units.

7. To make rectangles of width 1, adjust your slider for left rectangles to $n = 10$.

8. Press the *Show Sample Point* button. Point C moves along the tops of the rectangles. Try moving point C left and right.

9. Measure the y-coordinate of point C by selecting point C and then choosing **Ordinate (y)** from the Measure menu.

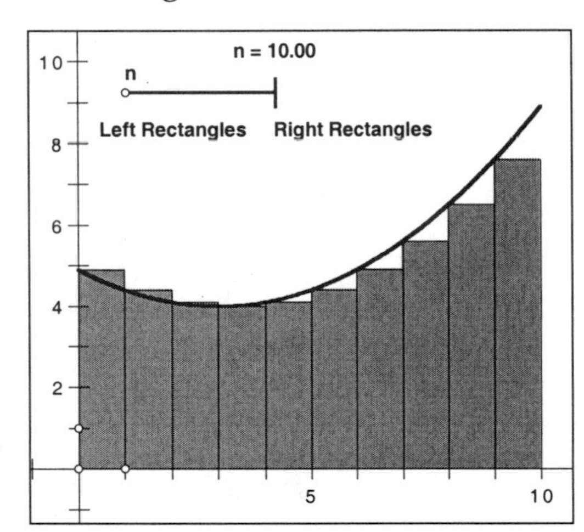

Your estimate should be the same as the value for *Area Sum*.

Q6 Use the y-coordinate of point C to calculate the area of each of the rectangles. Write down their areas and find the sum. What is your estimate for this case? Is it an underestimate or an overestimate?

10. Adjust your slider to show right rectangles with $n = 10$.

Save this answer to use in the Explore More Section.

Q7 Using right rectangles, what is the value for *Area Sum* in this case? Given just the expression for the function f, how could you find this sum? Is it an underestimate or an overestimate?

Using rectangles to estimate area, as you did above, will give you one overestimate and one underestimate. They can't both be overestimates or underestimates.

Q8 Back in steps 5–6, you looked at rectangles with a base of 10 units. One gave an underestimate of 49 square units, and the other gave an overestimate of 89 square units—a difference of 40 units. How far apart are your two estimates in Q6 and Q7 using rectangles with a base of 1 unit?

Q9 What do you think will happen to the difference between the estimates if you use rectangles of width 0.5?

11. Adjust the slider to show right rectangles with $n = 20$ and record the value of *Area Sum.* Then look at left rectangles with $n = 20$.

Q10 What is the difference between the estimates? Is it what you expected?

Did you find that increasing the number of subdivisions by a factor of 10 caused a reduction in the difference between the estimates by a factor of 10? Did doubling the number of subdivisions to 20 result in a further halving of the difference?

Q11 Use the slider to find the difference between the left and right rectangle sums for $n = 25$ or some other value. Explain how the difference between the estimates is related to the value of n. Can you explain why this relationship holds?

To see why this happens, press the *Next Page* button to go to page 5.

The sum of the areas of the rectangles shown when we're using the left endpoints to find the heights of the rectangles is called the *left sum.* When we use the right endpoints, it is called the *right sum.*

12. Experiment with the buttons to see just the left sum, then just the right sum, and then both sums at once.

13. While both sums are showing, adjust the slider so that $n = 1$. Then press the *Show Difference* button.

14. Check to see that when $n = 1$, you see a rectangle with an area of 40 square units—the difference between the areas of the two rectangles.

15. Adjust the slider so that $n = 4$. Note that the left and right sums have area in common. Press the *MovePointF* button to place the difference rectangle on top of the first left sum rectangle.

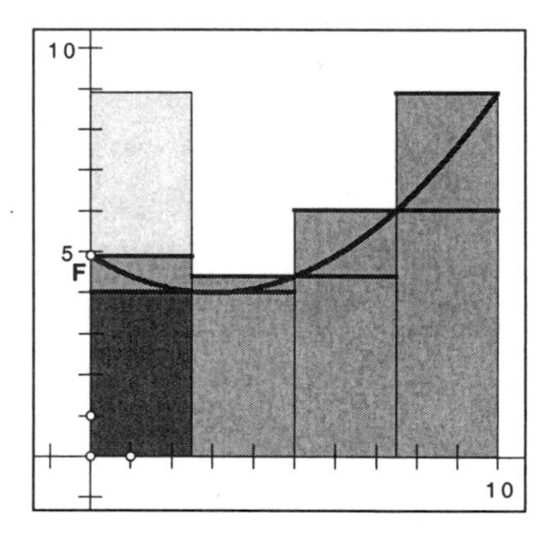

Q12 Increase the value of n. What do you notice about the height of the difference rectangle? Can you explain why this happens? What happens to the width of the rectangle?

Q13 Explain how you could use limits to show that you could make the difference between the left and right sum estimates as small as you wish.

Explore More

You can use limits to prove that the difference between the estimates approaches 0, but unfortunately, this does not tell you what area value the left and right sums approach. You will need a different limit for that.

1. Go to page 6. On this page is the linear function $f(x) = x + 1$.

Q1 What is the exact area of the shaded region under the curve $f(x) = x + 1$ on the domain $[0, 10]$?

Q2 Write an expression to calculate the area of the left sum with 5 rectangles.

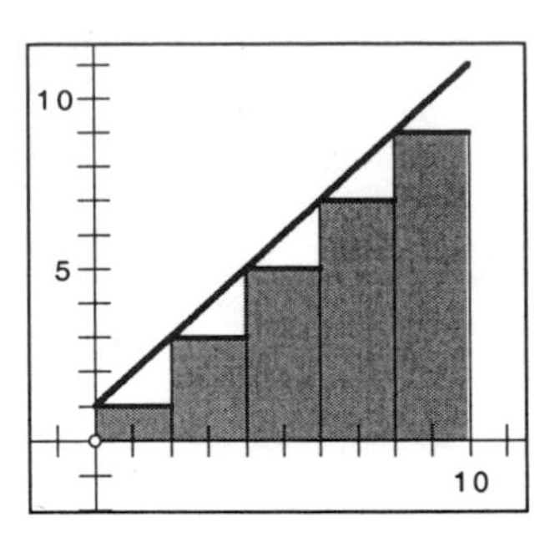

Q3 How would this expression change if there were 6, 7, 8, 10, 20, or even more rectangles? Write a general expression for the sum for n rectangles using Σ notation.

Q4 Using this expression, write a limit to express the area of the shaded region.

Q5 Write an expression for the sum of the 10 rectangles you made for the function $f(x) = 0.1(x - 3)^2 + 4$ back in Q7. How can you find the sum of these rectangles without adding everything up by hand?

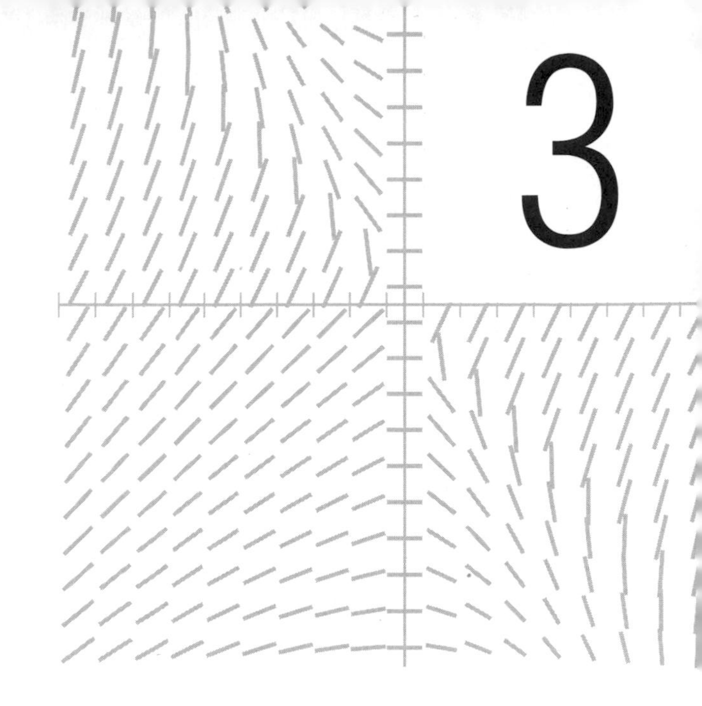

3

Exploring Derivatives

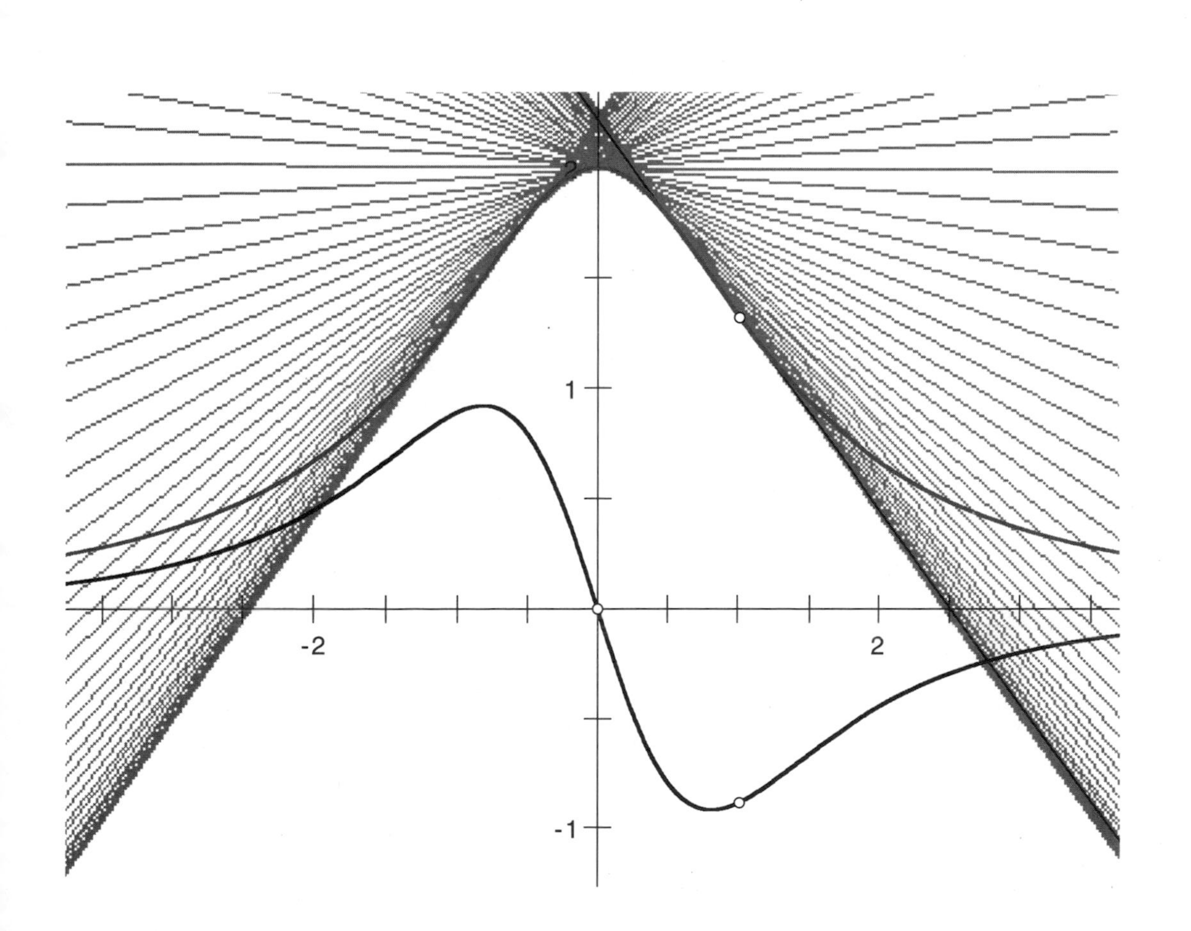

Taking It Near the Limit

Name(s): ______________________

If you want to model something in the real world with a function, like the gallons of gas your car uses per mile, it would be nice if the conditions always gave the same result—in this case, your mileage alone determines the number of gallons used. Unfortunately, the rate at which your car uses gas changes for all sorts of reasons—how fast you are going, your terrain, or even when you last changed your oil. A function is a good model, but we need something more. We need something that can predict and describe changing quantities as well as static ones.

The derivative of a function is such a tool. For a function $y = f(x)$, the *derivative* at a point is the instantaneous rate of change of f with respect to x at that point. In other words, the derivative tells you how much y is changing at any particular value of x. How do you find this derivative, and what does it look like? This activity will help you find out.

Sketch and Investigate

In order to start this process, you need a function.

1. **Open** the sketch **Derivative1.gsp** in the **Exploring Derivatives** folder. Choose **Plot New Function** from the Graph menu and enter $x^3 - 3x^2 + x$. Label this function f.

To construct a point, choose the **Point** tool and click on the function plot. Then with the **Text** tool, double-click on the point and type in your new label.

2. Construct two points anywhere on the function plot. Label these points P and Q.

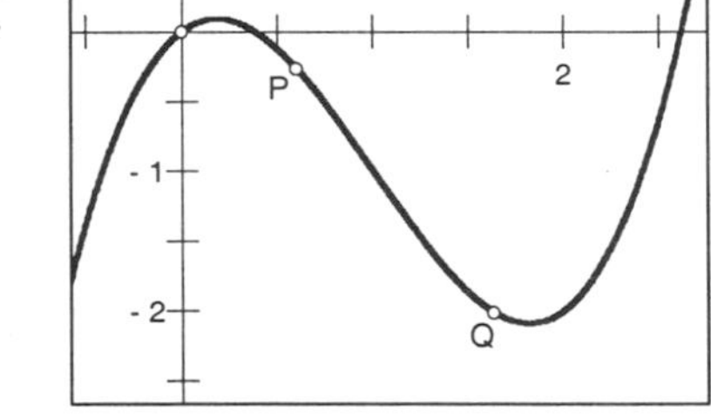

3. Select point P and point Q and measure their x-coordinates by choosing **Abscissa(x)** from the Measure menu.

4. Calculate $f(x_P)$ and $f(x_Q)$. Choose **Measure | Calculate.** Select the expression for f (not the plot!) to enter it into the calculator, and then select the measurement x_P. Repeat for point Q.

Recall that the average rate of change from $P\ (x_1, y_1)$ to $Q\ (x_2, y_2)$ is represented graphically as the slope of the secant line through P and Q:

$$rate_{avg} = \frac{\Delta y}{\Delta x} = \frac{y_2 - y_1}{x_2 - x_1} = slope(PQ)$$

Q1 In terms of the measurements you took in steps 3 and 4, what is the slope of the line PQ?

5. Construct the secant line PQ by selecting point P and point Q and choosing **Line** from the Construct menu.

Always select the measurement to enter into the calculator—not the point!

6. Find the average rate of change between the two points by calculating the slope of the line PQ. Choose **Measure | Calculate,** and then click

on each measurement to enter it into the calculator. Label this measurement *AverageRate PQ*.

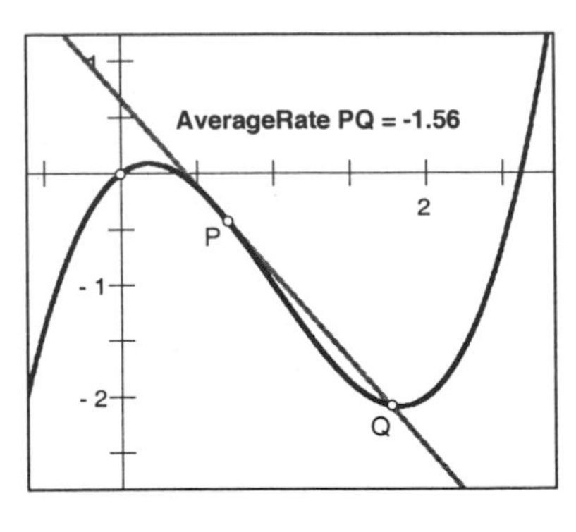

Your average rate of change (or secant slope) for f between point P and point Q tells you how much the function's y-value changes on the interval from x_P to x_Q. An instantaneous rate tells you how much the y-value is changing at a point—in this case, at point P. You can't use the slope formula with only one point (why not?) but you can look at the limit of the average rate or secant slope as point Q approaches point P.

7. Move point P so that its x-coordinate is as close to $x = 2$ as possible. You may not be able to get to $x_P = 2$ exactly, but that's okay.

8. Move point Q so that its x-coordinate is as close to $x = 2.1$ as possible.

9. Write down the x-value of 2.1 and the value of *AverageRate PQ* on your paper.

10. Move point Q so that its x-coordinate is as close to $x = 2.05$ as possible.

11. Write down the x-value of 2.05 and the value of *AverageRate PQ*.

Q2 From these two readings, what estimate would you give for the right-hand limit? If you place point Q as close as possible to $x = 1.9$ and $x = 1.95$, does the left-hand limit agree with the right-hand limit?

You may not be able to line them up exactly—if you can't, what *should* happen?

Q3 What happens to the slope measurement and the line if you actually put point Q in the same spot as point P? (After you do this, move point Q away from point P so they are distinct points again.)

You can zoom in at point P to get a better estimate for the limit or to check your estimate.

You can change a or b by double-clicking on the measurement and then typing in a new value.

12. Press the *Show Zoom Tools* button. The values a and b are the coordinates of the point you can zoom in on. The point P has coordinates (2, –2), so a should be set to 2 and b to –2.

13. Adjust the x- and y-scales at the same time by selecting points *x-scale* and *y-scale*, then click on either point and drag. Now when you move your mouse both sliders will move together. Try it.

As you zoom in, the axes should disappear. Press the *Show Side Axes* button. You can move these axes by dragging them.

14. Zoom in until your x-side-axis goes from approximately 1.2 to 2.4.

Although point Q will appear to move away from point P, note that the coordinates of point Q do not change. If you can, move point P closer to $x = 2$.

15. Move point Q so that its x-coordinate is as close to $x = 2$ as possible.

16. Write down the x-coordinate of point Q and the value of *AverageRate PQ* on your paper.

Q4 What estimate would you now give for the limit of the average rate or secant slope? (Do both sides still agree?)

If the tick marks on your axes disappear, zoom back out a little.

17. Zoom in one more time, but this time keep zooming in until your curve looks like a straight line. Again move point P as close to $x = 2$ as possible. (You should be able to be right at $x = 2.00000$.)

18. Move point Q so that its x-coordinate is as close to $x = 2$ as possible.

Q5 What estimate would you now give for the limit of the average rate or secant slope? Do both sides agree? Did it change much from your first initial estimate?

Q6 Around point P, can you tell the difference between a point on the line and a point on the curve?

You should have gotten a value of 1 for your limit of the average rate at $x = 2$.

$$\lim_{Q \to P} \textit{average rate from P to Q} = \textit{instantaneous rate at P} = \textit{the derivative at P}$$

so we write

$$\lim_{Q \to P} rate_{avg} = \lim_{Q \to P} \frac{y_Q - y_P}{x_Q - x_P} = \lim_{Q \to P} \frac{f(x_Q) - f(x_P)}{x_Q - x_P} = f'(x_P)$$

or simply $f'(2) = 1$. The notation f' stands for the instantaneous rate, or derivative.

Q7 You know that the graphical representation of the average rate of change from point P to point Q is the slope of the secant line PQ. Above, we've defined that

$$\lim_{Q \to P} \textit{average rate from P to Q} = \textit{instantaneous rate at P}$$

What do you think would be the graphical representation of the instantaneous rate (or derivative)? *Hint:* Zoom out (select both sliders and drag left) and look at the relationship between the function plot and the secant line. (Do not move point P or point Q.) What kind of line does the secant line look like now? We'll come back to Q7 and this idea in the next activity.

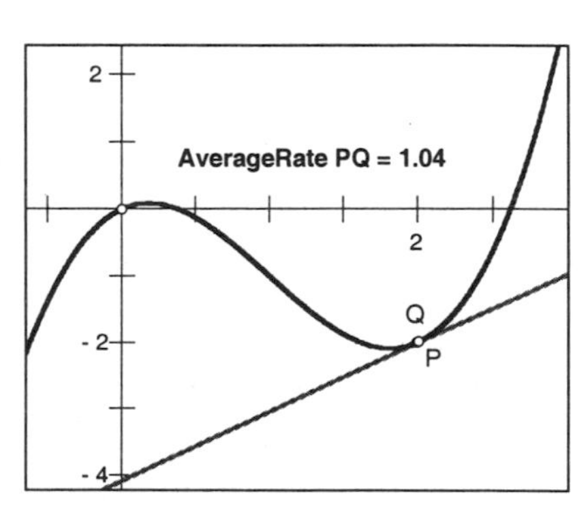

Now that you're zoomed out, you're ready to look for another instantaneous rate or derivative. An interesting place to look is at a point where the function changes direction.

19. Move point P as close as you can to the *local maximum point*—basically the top of a hill where f changes from going up to going down.

Before zooming in, be sure to change the value of a to the x-coordinate and b to the y-coordinate of the point where you want to zoom.

20. Move point Q as close as you can to point P. (See side note.) Zoom in a few times to get a good approximation for the x-coordinates of both the maximum and the limit.

Q8 What value for x_P gave you the biggest y-value or maximum? How can you be sure that your value for x_P is the best choice? What maximum value did you get?

Q9 What value did you get for the limit of the average rate for this point P?

Q10 For this point P, what is $\lim\limits_{Q \to P} \dfrac{f(x_Q) - f(x_P)}{x_Q - x_P}$, or $f'(x_P)$?

Q11 Use the sliders to zoom back out (but don't move point P or point Q). Looking at your secant line, what would you say about its slope?

Explore More

For each of these new functions and values, don't forget to change a and b to the center of where you want to zoom in!

1. Zoom out until your x-axis goes from approximately $x = -2$ to $x = 6$. Double-click on the expression for $f(x)$ and change the function to $f(x) = \sin(x)$. Click OK. Move point P so that its x-coordinate is as close to $x = \pi$ as possible. Repeat the above procedure to find your estimate of $\lim\limits_{x_Q \to \pi} \dfrac{f(x_Q) - f(\pi)}{x_Q - \pi}$, or $f'(\pi)$.

Q1 The above limit looks different from the first limit in the activity. How is the notation different from what you saw above? Has the change in notation altered the value of the limit or the average rate?

2. Zoom in on the following graphs at the following points, and make the value of x_Q as close as possible to x_P. (Don't forget to do this on both sides of point P.) You could run into some problems here.

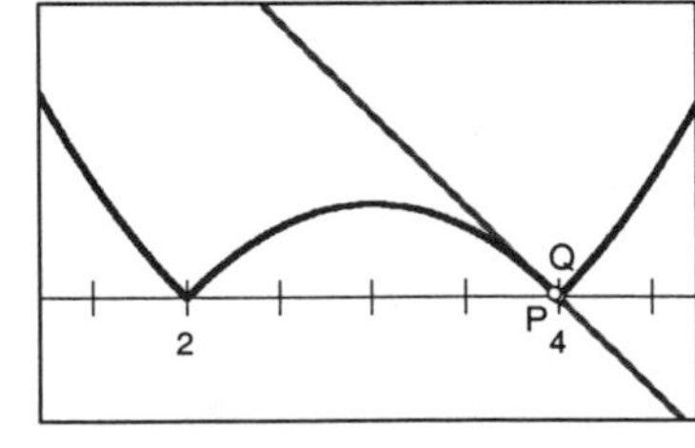

a. 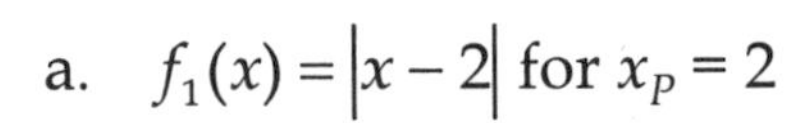 $f_1(x) = |x - 2|$ for $x_P = 2$

b. $f_2(x) = |x^2 - 6x + 8|$ for $x_P = 4$

c. $f_3(x) = \sqrt{x - 1}$ for $x_P = 1$

d. $f_4(x) = \ln(x)$ for $x_P = 1$

Q2 Are some of the cases in step 2 different from the previous examples? How? What conclusion can you reach? Justify your answers.

Going Off on a Tangent

Name(s): ______________________

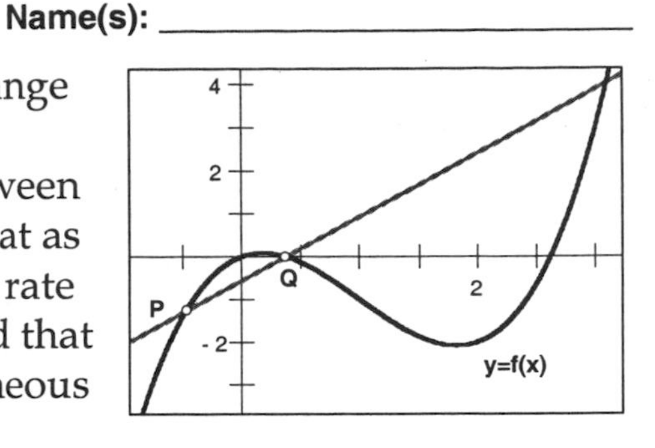

You can see what the average rate of change between two points on a function looks like—it's the slope of the secant line between the two points. You have also learned that as one point approaches the other, average rate approaches instantaneous rate (provided that the limit exists). But what does instantaneous rate *look like*? In this activity you will get more acquainted with the derivative and learn how to *see* it in the slope of a very special line.

Sketch and Investigate

1. **Open** the sketch **Tangents.gsp** in the **Exploring Derivatives** folder.

In this sketch there is a function plotted and a line that intersects the function at a point P. This new line is called the *tangent line* because it intersects the function only once in the region near point P. Its slope is the instantaneous rate of change—or derivative—at point P:

$$\textit{tangent's slope} = \textit{instantaneous rate at } P = f'(x_P)$$

So how do you find this line? Let's hold off on that for a bit and look at the line's slope—the derivative—and see how it behaves. Remember, slope is the key!

> Be careful here—the grid is not square!

Q1 Move point P as close as possible to $x = -1$. Without using the calculator, estimate $f'(-1)$—the derivative of f at $x = -1$. (*Hint:* What's the slope of the tangent line at $x = -1$?)

Q2 Move point P as close as possible to $x = 0$. Without using the calculator, estimate $f'(0)$—the derivative of f at $x = 0$. (*Hint:* See the previous hint!)

Q3 Move point P as close as possible to $x = 1$. Without using the calculator, estimate $f'(1)$—the derivative of f at $x = 1$. (Sorry, no hint this time.)

> If you'd like, you can animate point P by selecting it, then choosing **Animate Point** from the Display menu.

2. Move point P back to about $x = -1$. Drag point P slowly along the function f from left to right. Watch the line's slope carefully so that you can answer some questions.

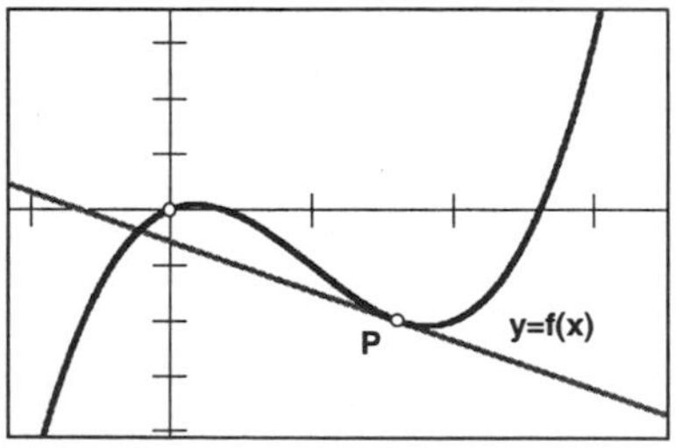

Q4 For what x-values is the derivative positive? (*Hint:* When is the slope of the tangent line positive?) What can you say about the curve where the derivative is positive?

Q5 For what x-values is the derivative negative? (*Hint:* Look at the hint in Q4 and make up your own hint.) What can you say about the curve where the derivative is negative?

Q6 For what x-values is the derivative 0? What can you say about the curve where the derivative is 0?

Q7 For what value or values of x on the interval from –1 to 3 is the slope of the tangent line the steepest (either positive or negative)? How would you translate this question into the language of derivatives?

3. Go to page 2 of the document. Here the function is $f(x) = 4\sin(x)$.

If you want to recenter your sketch, select the origin and move it to the desired location.

4. Press the *Show Zoom Tools* button and use the *x-scale* slider to change your window to go from -2π to 2π on the x-axis. (You can hide the tools again by pressing the *Hide Zoom Tools* button.)

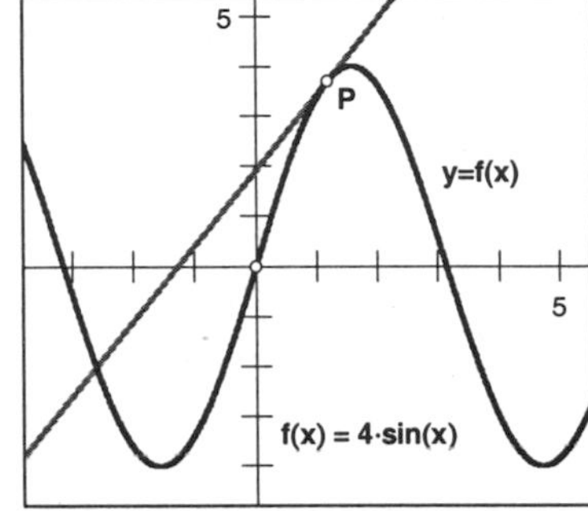

5. Move point P so that its x-coordinate is around $x = -6$.
6. Move point P slowly along the function to the right until you get to about $x = 6$. As you move the point, watch the tangent line's slope so you can answer the following questions.

Q8 Answer Q4–Q7 for this function. Could you have relied on physical features of the graph to answer these questions quickly? (In other words, could you have answered Q4–Q7 for this function without moving point P?)

There is an interesting relationship between how the slope is increasing or decreasing and whether the tangent line is above or below the curve. Move point P slowly from left to right again on the function, comparing the steepness of the line to its location—above or below the curve.

Q9 When is the slope of the line increasing? Is the tangent line above or below the function when the slope is increasing?

Q10 When is the slope of the line decreasing? Is the tangent line above or below the function when the slope is decreasing?

Q11 Write your conclusion for the relationship between the slope of the tangent line and its location above or below the curve. How would you translate this into a relationship between the derivative and the function's concavity?

Let's check whether or not your conclusion is really true. The derivative is the slope of the tangent line, so an easy way to check is to calculate the slope of the line.

7. Select the tangent line and measure its slope by choosing **Slope** from the Measure menu. Label it *TangentSlope*.

8. Move point *P* slowly along the function again from left to right and watch the values of the measurement *TangentSlope*.

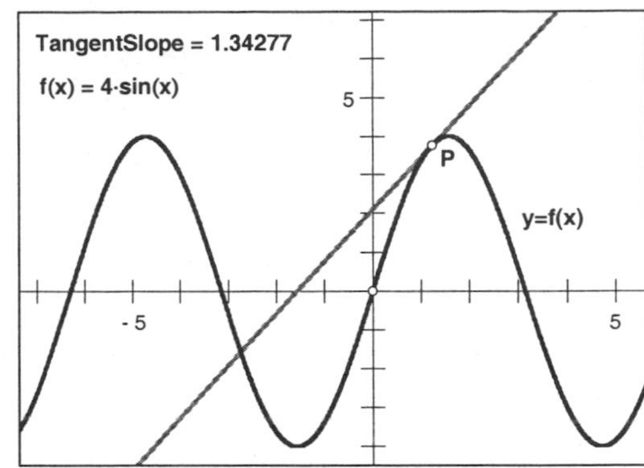

Q12 Do your answers to Q9–Q10 hold up?

Explore More

Each of the following functions has some interesting problems or characteristics. For each one, change the equation for $f(x)$ by double-clicking on the expression for $f(x)$ and entering in the new expression. Then answer the questions below. If you need to zoom in at a point, press the *Show Zoom Tools* button. Remember that *(a, b)* represents the point you will zoom in on. To change *a* or *b*, double-click on the parameter and enter a new value.

$$f_1(x) = |x-2|$$

$$f_2(x) = |x^2 - 6x + 8|$$

$$f_3(x) = \sqrt{x-1}$$

Q1 Where does the derivative not exist for $f_1(x)$ and why? (What happens to the tangent line at that point?)

Q2 Answer Q1 for $f_2(x) = |x^2 - 6x + 8|$.

Q3 Answer Q1 for $f_3(x) = \sqrt{x-1}$.

Q4 How is the function $f_1(x) = |x-2|$ different from all the others that you have looked at in this activity, including f_2 and f_3?

Plotting the Derivative

Name(s): ____________________

In this activity, you start with the plot of a cubic function, $f(x)$. Your job is to investigate the behavior of the slope of the function—its derivative as defined by the line tangent to the function. The goal of your investigation is to be able to predict and trace the resulting derivative function (the slope of the graph as a function of the x-values).

Sketch and Investigate

1. **Open** the document **PlotDerivative.gsp** in the **Exploring Derivatives** folder. Page 1 shows the plot of a cubic function. Sliders on the left allow you to vary the function.
2. Use the **Line** tool to construct a line with both construction points on the plot. Label the left point P and the right point Q with the **Text** tool.
3. With just point P selected, measure its x-coordinate by choosing **Abscissa (x)** from the Measure menu.

Select the line and choose **Slope** from the Measure menu.

4. Measure the slope of the line and move point Q relatively close to point P.

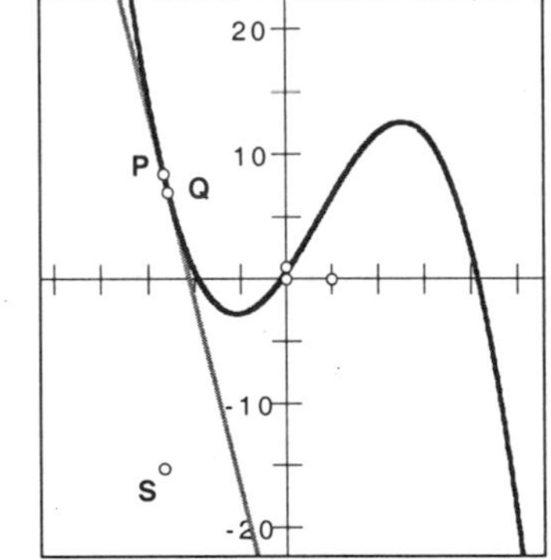

If the command is not enabled, make sure you have exactly two measurements selected, and nothing else.

5. Select x_P and the slope measurement, *slopePQ*, in that order. Choose **Plot As (x, y)** from the Graph menu. Label this point S.

Q1 *SlopePQ*, or the y-coordinate of point S, represents an approximation of the slope or derivative of $f(x)$ at point P. Explain why it is only an approximation and how you can minimize the approximation error.

To see the plot of the slope or "derivative" function, you can track the behavior of point S as points P and Q move along the graph.

6. With point S selected, choose **Display | Trace Plotted Point.**

If the shape is ragged, try again with the points closer together.

Q2 Using the **Arrow** tool, select just points P and Q, then click on either point and drag them slowly along the curve. What shape does the slope function trace out?

7. Erase the existing traces by pressing the Esc key (twice) or choosing **Erase Traces** from the Display menu.

You can speed them up by pressing the Up Arrow in the Motion Controller.

8. Your trace will be smoother if you animate the points rather than drag them by hand. With both points on the graph selected, choose **Animate Points** from the Display menu. Let the points go far enough to make a complete circuit back to their starting point. Don't press the *Reverse* button.

Q3 Do you notice any difference between the left-to-right trip and the right-to-left trip of the points? If so, why do you suppose this difference exists?

Q4 Press the *Pause* button in the Motion Controller, move the points farther apart. With points P and Q selected, press *Pause* again. Allow the motion to continue for a full cycle. What do you notice about the discrepancy between the two passes of the points?

A full cycle means going all the way to the right, then all the way to the left, and then back to your starting points.

Q5 Pause the animation, move points P and Q as close together as you can, and erase your traces. Again select both points and release the *Pause* button. Allow the motion to continue for a full cycle, then stop. What do you notice this time about the discrepancy between the two passes of the points? (Don't erase the trace this time.)

By using two arbitrary points, you have plotted point S with a y-value of $\dfrac{f(x_Q)-f(x_P)}{x_Q-x_P}$. By taking point Q close to point P, your y-value approximates $\lim\limits_{Q\to P}\dfrac{f(x_Q)-f(x_P)}{x_Q-x_P}$, or the derivative at $x = x_P$. Now, if you fix point Q's x-coordinate a set distance, h, from point P's x-coordinate, you'll still be approximating the derivative. You'll just be using another form of the definition and approximating $\lim\limits_{h\to 0}\dfrac{f(x_P+h)-f(x_P)}{h}$ instead.

9. Select point Q and choose **Split Point From Function Plot** from the Edit menu. (Still, don't erase the trace.)

Click on a measurement or function to enter it into the calculator.

10. Choose **Calculate** from the Measure menu and calculate $x_P + h$. Then calculate $f(x_P + h)$.
11. Plot the point $(x_P + h, f(x_P + h))$ by selecting $x_P + h$, then selecting $f(x_P + h)$, and choosing **Plot as (x, y)** from the Graph menu.

Use the **Text** tool to relabel the merged point Q.

12. Select point Q and the new point and choose **Edit | Merge Points.**

Now you have a secant line and plotted point similar to the ones you constructed by hand in steps 1–6, but with the value h determining the horizontal separation between the two points on the function. This allows you to animate just point P and know that the points on the function will remain separated by the same amount. It also allows you to actually construct the locus of the plotted point S.

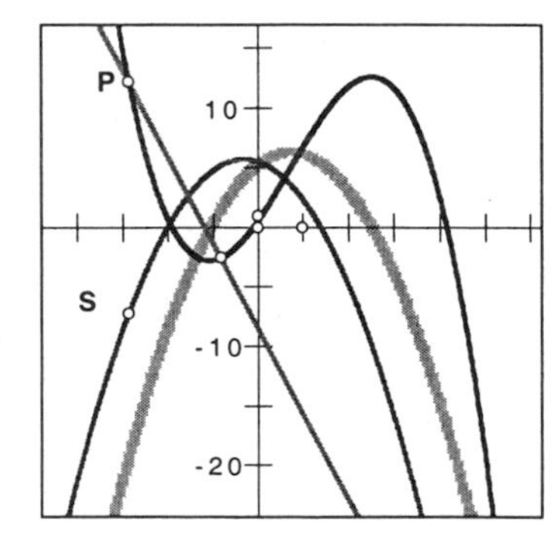

13. Turn off tracing for point S (but don't erase the previous trace). With point S still selected, select point P and choose **Construct | Locus.**

14. Experiment with the slider for h to see how it affects the locus of point S.

Q6 What must you do with your slider for h to make the locus match the trace you made in Q5? Explain what you are doing mathematically when you match up the trace and the locus by adjusting h. (When you are done, press the Esc key twice to erase the trace.)

Explore More

In the activity "Going Off on a Tangent," you used the slope of the tangent line to find the derivative at a point. Could you make a rough sketch of the derivative by tracking how the tangent line's slope changes?

1. Select the locus and point S, then choose **Hide/Show** from the Action Buttons submenu in the Edit menu.

2. Press the *Hide Objects* button. Then adjust any of the sliders a–d to make a new cubic function.

Check the status line or the Motion Controller to see if point P is selected. If it isn't, click on the point again.

3. Adjust the h slider as close as possible to $h = 0$ so that you have a good approximation of the tangent line at $x = x_P$. Move point P along the function plot and observe how the slope of this approximate tangent line changes. When is the slope of the line positive? Negative? Zero?

4. Predict what you think the locus of point S will look like and make a little sketch of your prediction in the margin.

5. Press the *Show Objects* button. Compare your prediction with the locus shown. How did you do?

By comparing the slope of the secant line (approximated tangent line) with the location of point S as you drag point P, you can discover the relationship between them. This will help you predict what any function's derivative plot will look like. The following questions will help you discover these relationships. To answer these questions, move point P along the function and focus on the slope of the line and the location of point S.

Q1 When the slope of the line is positive, what can you say about the location of point S? When the slope of the line is negative? When the slope is 0?

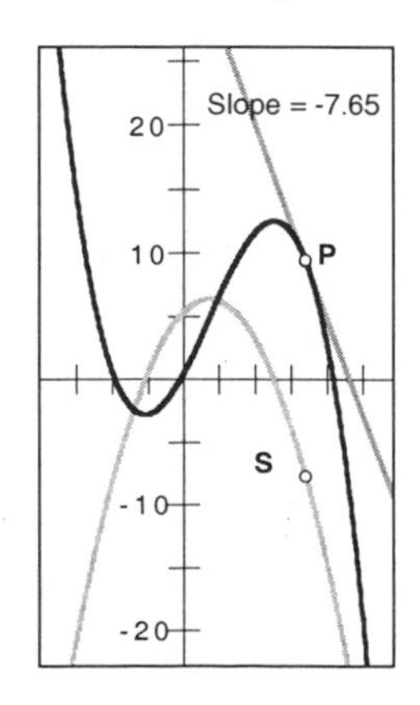

Make sure you are looking at the slope of the line and not the function f.

Q2 Drag P slowly from left to right. When the *slope* of the line is increasing, what can you say about the location of point S? When the slope of the line is decreasing?

6. Check your answers by pressing the *Hide Objects* button and adjusting the sliders to create a different function plot. Move point P along the function,

make a prediction for the locus, and sketch it in the margin. Then show the locus and see how you did this time.

7. Once you are satisfied you can predict the graph for these cubic functions, double-click on the expression for $f(x)$ and edit it to $f(x) = ax^4 + bx^3 + cx^2 + dx + e$. Repeat step 4.

8. Adjust the sliders to create a variety of fourth degree polynomial functions. (Make sure to try some negative values for the sliders.) Try to predict the locus in each case.

Q3 Describe the graph that point S sweeps out. Is it consistent with your answers to Q1 and Q2?

Try predicting what the locus will look like for other functions (see below), and then showing the locus to check your predictions.

For each function, describe the locus traced by point S for the functions you have created and check that they are consistent with your answers to Q1 and Q2.

Examples to try:

$f_1(x) = a^x$ for $a > 1$

$f_2(x) = a^x$ for $0 < a < 1$

$f_3(x) = a\sin\ (bx)$

$f_4(x) = \dfrac{a}{x^2 + b}$

$f_5(x) = \sqrt{9 - x^2}$

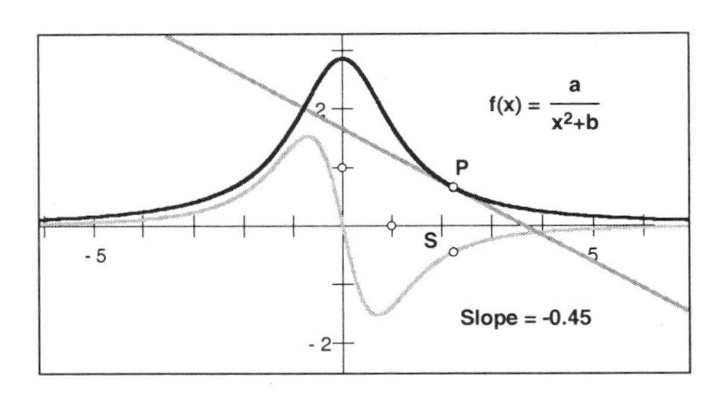

So What's the Function?

Name(s): ______________________

In earlier activities, you created traces or plots of the approximate derivative by finding the slope of the tangent line at a point and then plotting the slope, one point at a time. It would be so much easier and faster if we could create a function that calculates the derivative at every point all at once. Sound impossible? Luckily it isn't, and that is what this activity is all about.

Sketch and Investigate

The derivative of $f(x)$ at $x = x_P$ can be approximated by the ratio $\frac{f(x_P + h) - f(x_P)}{h}$ for h near 0. We can easily do the same thing to approximate the derivative *function*—just use the variable x in place of the constant x_P.

1. **Open** the document **DerFunction.gsp** in the **Exploring Derivatives** folder. On page 1 you will find a linear function $f(x) = bx + c$, its plot, and various sliders.

Q1 You know that the derivative at a point is the function's slope at that point, so for a linear function, what is the derivative at all points?

To check your answer graphically. . .

2. Choose **Plot New Function** from the Graph menu and for $g(x)$ enter the expression $(f(x + h) - f(x))/h$.

3. Using the **Point** tool, plot a point anywhere on the function $f(x)$. Label this new point P.

Click on the expression for g and then on measurement x_P to enter them into the calculator.

4. With point P selected, measure the x-coordinate for point P by choosing **Measure | Abscissa (x)**. Then choose **Calculate** from the Measure menu and calculate $g(x_P)$.

Select the measurements x_P and $g(x_P)$, in that order. Then choose **Plot As (x, y)** from the Graph menu.

5. Plot the point $(x_P, g(x_P))$. Label this new point S with the **Text** tool.

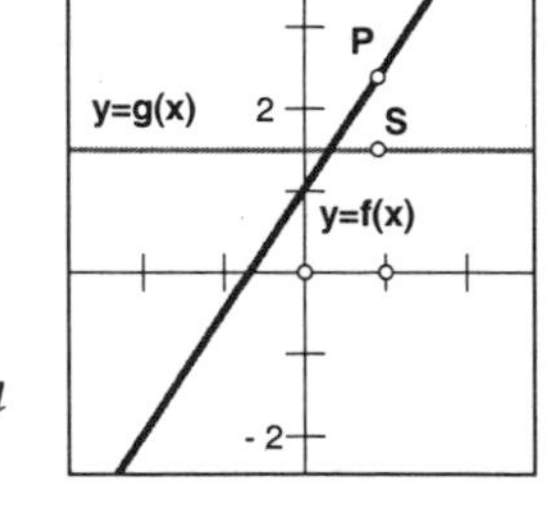

6. Experiment with the sliders for b, c, and h to see how they affect the y-coordinate of point S.

Q2 Show why $g(x)$ is the horizontal line $y = b$ for any h and c by simplifying the expression in step 2 algebraically for $f(x) = bx + c$.

7. Edit $f(x)$ by double-clicking on the expression for f, deleting $bx + c$, and entering $f(x) = ax^2$.

You can also use the button provided to set $h = 0.00001$.

8. Adjust the h slider so that you have a very good approximation for the derivative function.

So What's the Function? (continued)

Q3 Experiment with the a slider; what *kind* of function do the derivative approximations for $f(x) = ax^2$ appear to be?

Q4 Press the *Case: a = 1* button. What function does the derivative approximation function appear to be if $a = 1$? (Write an equation.)

Q5 Examine the values of x_P and $g(x_P)$ as you move point P along the function plot of $f(x)$. What is the relationship between x_P and $g(x_P)$? Do your observations support your answer to Q4?

Q6 Press the other *Case* buttons. In each case, make a table and write down at least three coordinate pairs $(x_P, g(x_P))$. Try to find a relationship between the x-coordinate, x_P, and the y-coordinate, $g(x_P)$, of the derivative approximation. Write your function for $g(x)$. Does your function agree with the basic shape you gave in Q3?

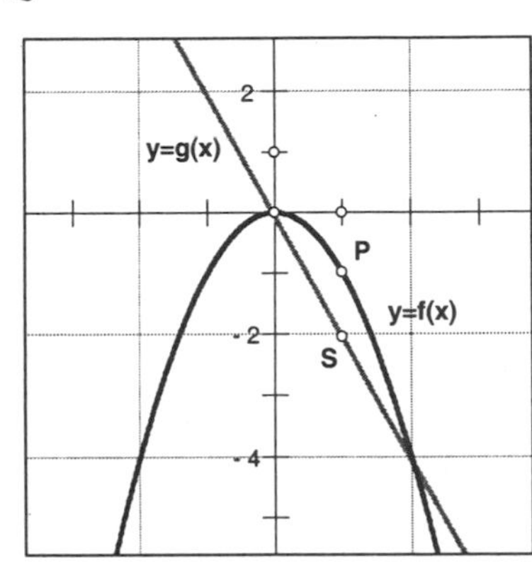

To graphically verify your results, you can plot your function from Q6. Make sure your function matches the function g for *all* values of a!

Q7 Algebraically verify your results in Q6 by simplifying the expression in step 2 for $f(x) = ax^2$ and then calculating its limit as h approaches 0.

Q8 Putting together your work from Q2 and Q7, what do you think the derivative function will be for $f(x) = ax^2 + bx + c$? Check your answer by editing $f(x)$ and then pressing the *Show Derivative* button.

Q9 Experiment with the sliders a, b, and c one at a time. For each one, describe how it affects the plot of f and then how it affects the plot of the derivative of f.

9. Press the *Hide Derivative* button. Then edit $f(x)$ to $ax^3 + bx^2 + cx + d$.

10. Experiment with the sliders that define $f(x)$. Make sure that the h slider is still in the position to give a very good approximation for the derivative function.

Q10 What *kind* of function does the approximation for the derivative appear to be for a cubic function?

11. Press the *Case: a = 1* button to set $a = 1$ and the rest of the sliders to 0. Examine the values of x_P and $g(x_P)$ as you move point P along the plot of $f(x)$.

Q11 Write down at least three coordinate pairs $(x_P, g(x_P))$ and find a relationship between the x-coordinate, x_P, and the y-coordinate, $g(x_P)$, of the derivative approximation. What function does the derivative approximation function appear to be if $a = 1$? (Write an equation.)

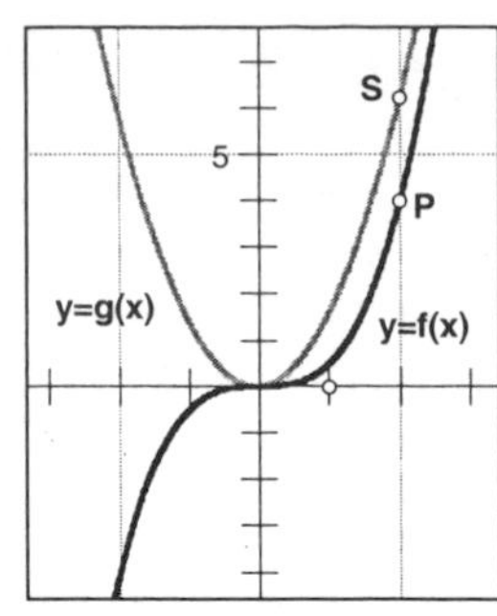

Q12 Press the other *Case* buttons. In each case, make a table and write down at least three coordinate pairs $(x_P, g(x_P))$ to find a relationship between the x-coordinate, x_P, and the y-coordinate, $g(x_P)$, of the derivative approximation. Write your function for $g(x)$. Does your function agree with the basic shape you gave in Q10?

Q13 Algebraically verify your results from Q12 by simplifying the expression in step 2 for $f(x) = ax^3$ and calculating its limit as h approaches 0.

Q14 Putting together your work from Q2, Q7, and Q13, what do you think the derivative function will be for $f(x) = ax^3 + bx^2 + cx + d$? Check your answer by editing $f(x)$ and then pressing the *Show Derivative* button.

Q15 Experiment with the sliders a, b, c, and d one at a time. For each one, describe how it affects the plot of f and then how it affects the plot of the derivative of f.

Exploration 1

Finding the derivative algebraically by using the limit definition is much easier for polynomials than for most other functions, such as logarithms, trigonometrics, exponentials, roots, and so on. But the graphical procedure for approximating them is much the same.

1. Edit $f(x)$ to one of the basic functions given below.
2. Make a conjecture about its derivative function by examining the basic shape of the function $g(x)$. Then use the relationship between the x-coordinate, x_P, and the y-coordinate, $g(x_P)$, to predict or figure out the exact equation.
3. Write your function for $g(x)$ and check by pressing the *Show Derivative* button.

Basic Functions:

$$f(x) = \sin(x), \cos(x), \text{ or } \tan(x)$$

$$f(x) = \sqrt{x} \text{ or } x^{(p/q)}$$

$$f(x) = \frac{1}{x} \text{ or } \frac{1}{x^2}$$

$$f(x) = \log(x) \text{ or } \ln(x)$$

$$f(x) = a^x \text{ for } 0 < a < 1 \text{ or for } a > 1$$

Exploration 2

When we were looking for the derivative for the general quadratic and cubic functions, you used your previous work and added the two derivatives together—in other words,

$$(ax^2 + bx + c)' = (ax^2)' + (bx + c)'$$

Is it always true that $(p_1 + p_2)' = (p_1)' + (p_2)'$, for any functions p_1 and p_2?

To label the function p_1, double-click on the expression for $p_1(x)$ with your **Text** tool and label it p[1].

1. Go to page 2. Choose **New Function** from the Graph menu to create a new function $p_1(x) = \sin(x)$. Then create another new function $p_2(x) = x$. Do not plot these functions.
2. Edit $f(x)$ to $p_1(x) + p_2(x)$.
3. Calculate the derivative for $p_1(x)$ by selecting it and choosing **Derivative** from the Graph menu.
4. Calculate the derivative for $p_2(x)$.

Click in an empty spot to deselect all objects before you go to the Graph menu.

5. Create a new function $r(x) = p_1'(x) + p_2'(x)$ by this time choosing **Plot New Function** from the Graph menu. Make the resulting plot a new color.
6. Adjust the slider for h so that $g(x)$ is a very good approximation for the derivative of $f(x)$.

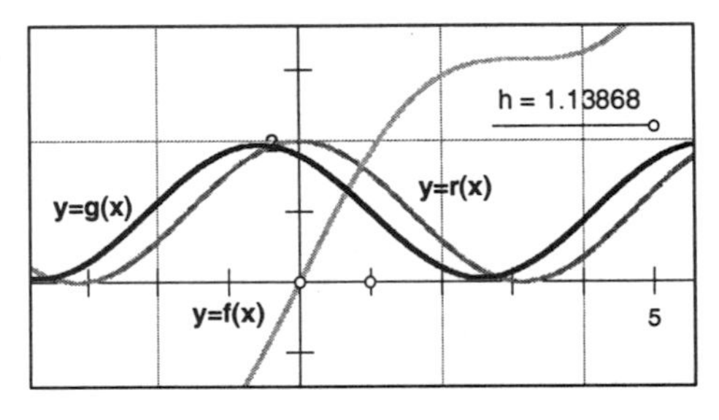

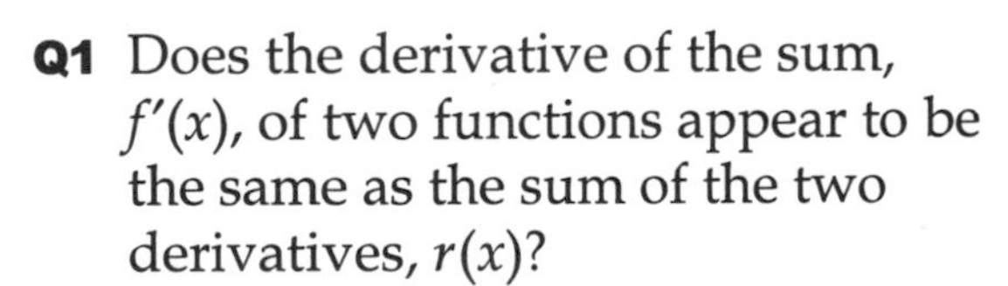

Q1 Does the derivative of the sum, $f'(x)$, of two functions appear to be the same as the sum of the two derivatives, $r(x)$?

Investigate other combinations of p_1 and p_2 by editing $f(x)$ and $r(x)$ to answer the following questions:

Q2 Does the derivative of the difference $f'(x)$ of two functions, $f(x) = p_1(x) - p_2(x)$, appear to be the same as the difference of the two derivatives, $r(x) = p_1'(x) - p_2'(x)$?

Q3 Does the derivative of the product $f'(x)$ of two functions, $f(x) = p_1(x) \cdot p_2(x)$, appear to be the same as the product of the two derivatives, $r(x) = p_1'(x) \cdot p_2'(x)$?

Q4 Does the derivative of the quotient $f'(x)$ of two functions, $f(x) = p_1(x)/p_2(x)$, appear to be the same as the quotient of the two derivatives, $r(x) = p_1'(x)/p_2'(x)$?

Derivatives of Exponential Functions

Name(s):______________________________

Consider the family of functions, $f(x) = a^x$ for $a \geq 0$. Does the derivative of f end up being a log or an exponential or does it follow the power rule? In this activity, you will explore these possibilities and see if you can derive a formula for the derivative of an exponential function.

Sketch and Investigate

Start with a doubling function, $f(x) = 2^x$. Sketch its basic graph by hand. Is the graph increasing or decreasing? Concave up or concave down? Draw a rough sketch of what the derivative should look like based on your basic graph's behavior. Any guesses yet on what the derivative might be? Can you eliminate any choices?

This time, instead of sketching a tangent line, you will use the definition of the derivative, $\lim_{h \to 0} \frac{f(x+h) - f(x)}{h}$, and approximate this limit using Sketchpad, thus creating an approximation for the derivative of $f(x) = 2^x$.

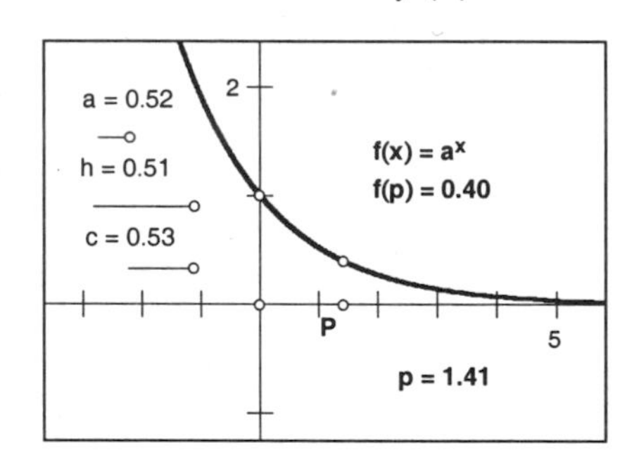

1. **Open** the sketch **Exponent.gsp** in the **Exploring Derivatives** folder. Axes and a point P $(p, 0)$ are included, as are three sliders labeled a, c, and h. The measurement p is the x-coordinate of point P.

Choose **Plot New Function** from the Graph menu.

2. Plot a new function $f(x) = a^x$.

Calculate $f(p)$, then select p and $f(p)$, in that order, and choose **Plot as (x, y)** from the Graph menu.

3. Plot the point that corresponds with $x = p$ on $f(x)$.
4. Drag a and observe the behavior of the function for different values of a.

Q1 For the family of functions $\{f(x) = a^x \mid a \geq 0\}$, what values for a give increasing functions? Decreasing functions?

Q2 What happens when $a = 1$? Why? What happens when $a \leq 0$? Why?

Q3 Compare the graph of $f(x) = a^x$ for two values of $a > 1$. How are the graphs the same? How are they different?

Q4 Using the limit definition of the derivative, what is the expression for the derivative of $f(x) = 2^x$?

How can you use Sketchpad to model this limit and derivative? (*Hint:* The sketch has more sliders that you have not used. No fair peeking.)

Make g a new color. To label your graphs, click on the plot once with your **Text** tool.

5. Plot $g(x) = \dfrac{a^{(x+h)} - a^x}{h}$.

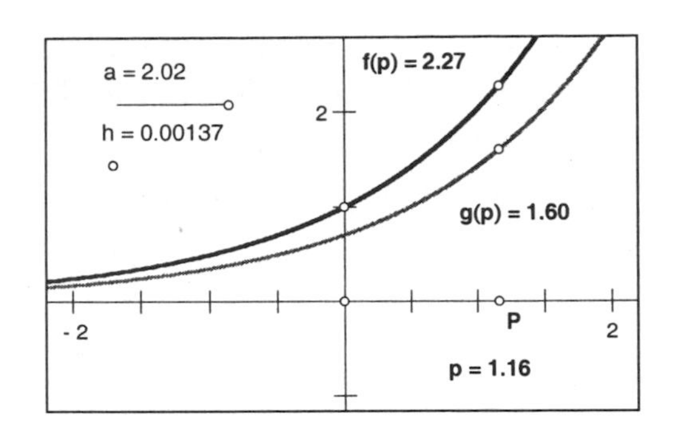

6. Set the slider for a as close as possible to 2.
7. Set the slider for h as close as you can to 0.

Calculate $g(p)$. Select p and $g(p)$ in that order, then choose **Plot as (x, y)** from the Graph menu.

8. Plot the point that corresponds with $x = p$ on $g(x)$.

Q5 Why is $g(x)$ not the derivative of $f(x) = a^x$? What does $g(x)$ represent?

Q6 How can you use $g(x)$ and Sketchpad to find a good approximation for $f'(x)$?

Q7 Compare $g(x)$ to your basic graphs—what type of function is g? Does g look like the rough sketch you did at the beginning?

The graphs you have created can be altered by varying the slider for a. Before dragging a, predict what you will see for the "derivative" for various values of a: $a \leq 0$, $0 < a < 1$, $a = 1$, and $a > 1$. Once you have your prediction, drag a through various values and see if your predictions are correct.

Q8 What happens when $a = 1$? $a \leq 0$?

Q9 What happens to our "derivative" when $0 < a < 1$? Why?

Q10 What happens when you increase a to values greater than 1?

In the sketches below, a has been changed until the graphs of f and g appear to coincide. Using the slider for a, figure out when this happens.

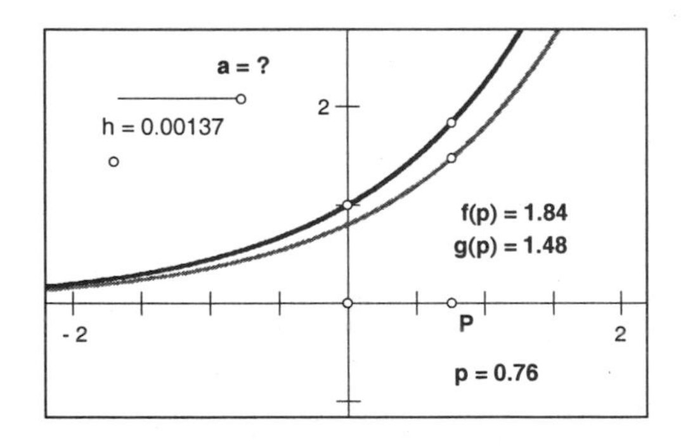

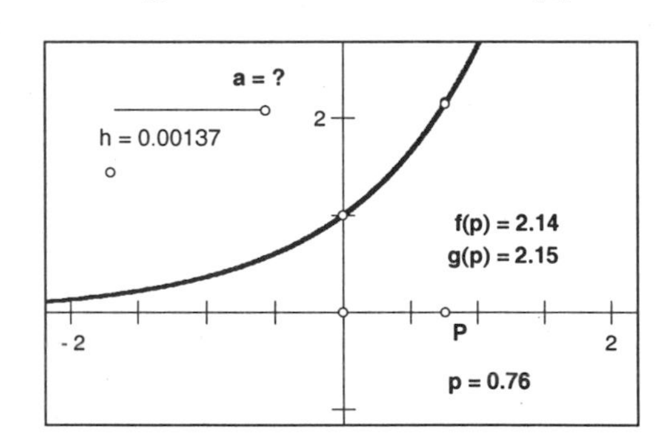

Q11 What value of a makes f and g coincide?

That the two plots can coincide suggests that a function can be its own derivative! The number a at which this occurs appears to be close to the value of the number $e \approx 2.7183$. Is e the value of a for which the function $f(x) = a^x$ is its own derivative? Sketchpad cannot prove this to be true, but it can be used to explore the possibility.

Vary a again. When the graphs of $f(x) = a^x$ and its derivative do not coincide, how do they appear to be related? This time, focus on a possible equation for the graph of the derivative of $f(x) = a^x$. Remember there are translations: $f(x + c)$ and $f(x) + c$, and stretches: $f(cx)$ and $cf(x)$.

This is a hard question, so let's experiment with f and a new slider, c.

Choose **Plot New Function** from the Graph menu. To enter f and c in the calculator, select them with the **Arrow** tool.

9. Plot a new function, $h(x) = f(x + c)$, and hide the graph of f by selecting just the plot of $f(x)$ and choosing **Edit | Action Buttons | Hide/Show.** Then press this new button to hide f.

10. Fix any $a > 1$ and vary c. Can you get your new graph to coincide with the graph of the "derivative"? If so, continue with step 11. If not, go to your next transformation in step 12.

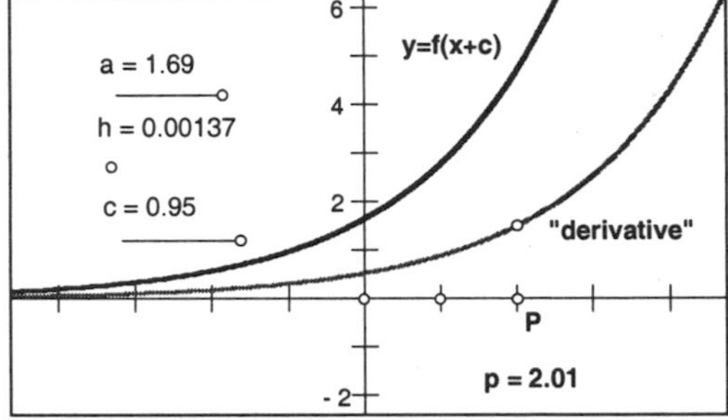

11. Fix any $0 < a < 1$ and vary c. Can you get your new graph to coincide again? If so, then you've found the right transformation. If not, let's go to the next one.

12. Double-click on the transformation expression $h(x) = f(x + c)$—not the plot of h—and change the expression to $f(x) + c$, $f(cx)$, or $cf(x)$.

13. Repeat steps 10–12 until you discover the transformation that works for both intervals: $a > 1$ and $a < 1$.

When you have discovered which transformation works, you will want to find a value for c. What value of c is needed to create the sketch below when $a \approx 2$? Is there any relationship between the value of c and the number 2?

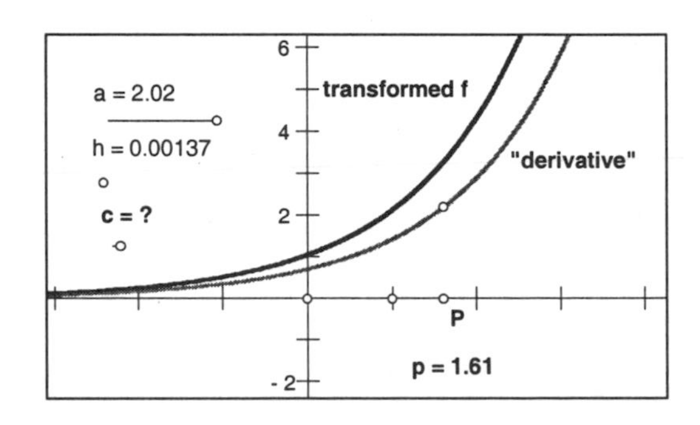

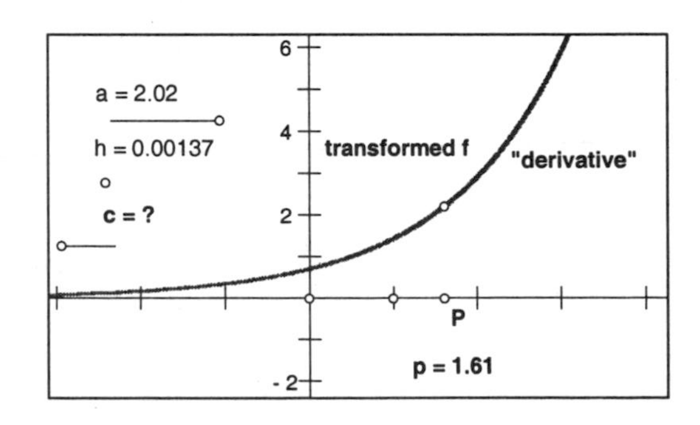

Explore More

Did you discover that the transformation is a vertical stretch, so $f'(x) \approx g(x) \approx c \cdot f(x)$? If you didn't, try steps 9–13 again. If you divide both sides of $g(x) \approx c \cdot f(x)$ by $f(x)$ you end up with $\dfrac{g(x)}{f(x)} \approx c$. But c is a constant! So the ratio of the y-values is a constant for each value of a. (*Note:* This is true for any vertical translation.) In the next few steps, you will use Sketchpad to check this fact. This may also lead to a formula for the derivative of the general exponential function, $f(x) = a^x$.

1. Use the calculator to find the ratio of $\frac{a^{(x+h)} - a^x}{h}$ to a^x by using the ratio $\frac{g(p)}{f(p)}$.
2. Relabel this ratio *constant*.
3. Fix any $a > 0$. Slide point P along the x-axis to change the value of p.

Q1 Does the measurement *constant* change as the value of p changes?

4. Fix point P. Change the value of a using the slider.

Q2 Does the measurement *constant* change as the value of a changes?

It's hard to discern a pattern by just looking at a bunch of numbers. Remember, you are trying to figure out the value for *constant* for different values of a.

So try plotting the values.

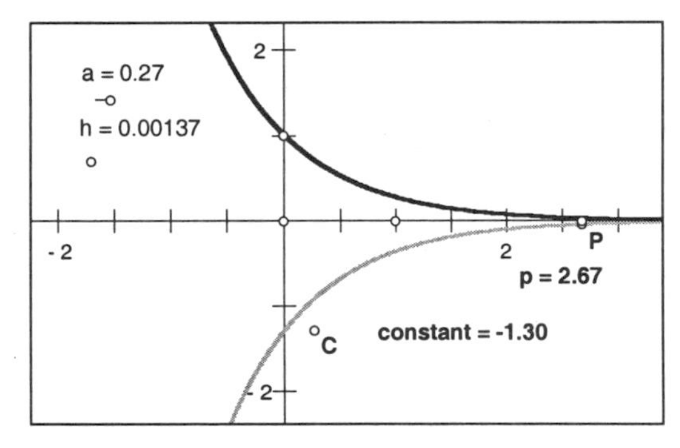

5. Select the measurement a and the measurement *constant*. Choose **Plot as (x, y)** from the Graph menu.
6. With just your new point selected, choose **Trace Plotted Point** from the Display menu. Label the plotted point C.

Q3 Fix any $a > 0$. Slide point P along the x-axis to change the value of p. Does point C move as point P changes?

Q4 Fix point P. Change the value of a using the slider. Does point C move as point P changes?

To see the locus of points, you can select point *C* and point *a*—the actual point on the end of slider *a*—in that order. Then choose **Locus** from the Construct menu.

Q5 One of the above movements created a trace or locus of points. Do you recognize the graph? Any ideas now what the derivative of $f(x) = a^x$ is?

Derivatives and Transformations

Name(s): ______________________

If you know the derivative for a given basic function, can you predict the derivative for a transformation of that basic function? For example, the derivative for $f(x) = \sin(x)$ is $f'(x) = \cos(x)$, so what is the derivative for $f(x) = 2\sin(3x) + 4$? In this activity, you will explore this question graphically.

Sketch and Investigate

There are four basic possible transformations: vertical and horizontal translations, and vertical and horizontal stretches. If you have a basic function $f(x)$, you can represent all of them at once by $g(x) = af(bx - c) + d$. We'll investigate these transformations one at a time to see how each constant affects the plot graphically and how it affects the derivative.

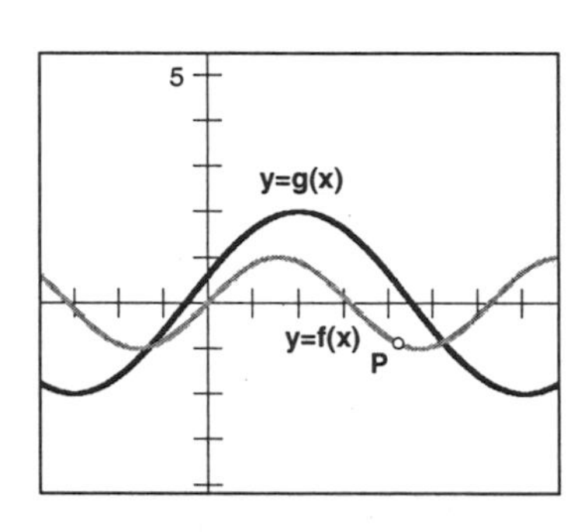

1. **Open** the document **Transformations.gsp** in the **Exploring Derivatives** folder. On page 1 is the function $f(x) = \sin(x)$ and the transformation $g(x) = f(x) + d$.

2. Adjust the slider for d.

Q1 What kind of graphical transformation occurs when you change the value of d? Specifically, what happens if d is negative? Positive? Why does this occur?

Q2 How do you think the derivative is affected by the transformation?

One way to see how the derivative is affected is to examine the slopes of the tangent lines at corresponding points—points P_f and P_g.

3. Press the *Show Tangents* button.

You can animate point P_f instead by selecting point P_f and choosing **Animate Point** from the Display menu.

4. Move point P_f along the function $f(x)$.

Q3 How is the slope of the tangent line to the function $g(x)$ at point P_g related to the slope of the tangent line to the function $f(x)$ at point P_f?

Q4 Write an equation for $g'(x)$ in terms of $f'(x)$.

Select the expression for f' to enter it into the New Function panel.

5. To check your answer to Q4, press the *Show* f' button and choose **Plot New Function** from the Graph menu. Enter your equation from Q4. Then press the *Show* g' button.

Q5 Is the function you plotted in step 5 identical to the plot of g'? If so, great! If not, try again.

Adding d to the function's value is one type of transformation. What happens if a constant is added to, or subtracted from, the x-value first?

6. Go to page 2 of the document. Here we have the transformation $g(x) = f(x - c)$.
7. Adjust the slider for c.

Q6 What kind of graphical transformation occurs when you change the value of c? Specifically, what happens if c is negative? Positive? Why does this occur?

Q7 How do you think the derivative is affected by the transformation?

You can also see how the derivative is affected by tracing a point representing the slope of the secant line between the two points $(x, g(x))$ and $(x + h, g(x + h))$, and then having h approach 0.

8. Choose **Secant Line** from **Custom** tools. Click on point R (on the plot of function g), and then on the expression for $g(x)$.

This tool makes the new point $(x_R + h, g(x_R + h))$ and calculates the measurement *SlopeSecant,* which is the value $\dfrac{g(x_R + h) - g(x_R)}{h}$.

9. Measure the x-coordinate of point R. With the x-coordinate selected, select the measurement *Slope Secant* and choose **Plot as (x, y)** from the Graph menu. Label this new point S. Make it a new color.
10. Press the *h→0.00001* button and then turn on tracing for point S. With $h = 0.00001$, the trace of point S will be a very good approximation of $g'(x)$.

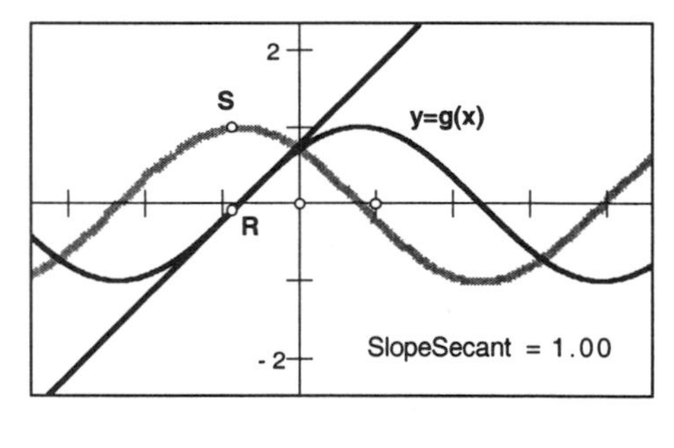

If you want to eliminate clutter, you can press the *Hide f* button here.

11. Select point R and choose **Animate Point** from the Display menu.

Q8 Press the *Show f′* buttons. Compare your approximations of $g'(x)$ and $f'(x)$.

Q9 Write an equation for $g'(x)$ in terms of $f'(x)$.

12. To check your answer to Q9, choose **Graph | Plot New Function** and enter your equation from Q9. Press the *Show g′* button.

If you don't see a new function when you press the *Show g′* button, then that means the two functions are identical and you are right!

Q10 Is the function you plotted in step 12 identical to the plot of $g'(x)$? If so, great! If not, there are many transformations of the derivative you can try: $f'(x)$, $f'(x) + k$, $f'(x + k)$, $f'(kx)$, $k \cdot f'(x)$, and $k \cdot f'(kx)$. Adjust the slider for k to make your prediction function match your trace. If you can't, then it is time to try the next transformation.

We've now looked at all the transformations possible by adding or subtracting a constant. What kind of transformations occur when you multiply by a constant?

13. Go to page 3 of the document. Here we have the transformation $g(x) = a \cdot f(x)$, with points S and R defined as they were on page 2.
14. Adjust the slider for a.

Q11 What kind of graphical transformation occurs when you change the value of a? Specifically, what happens when a is negative? Positive? Why does this occur?

Q12 How do the graphs compare when a is greater than 1? Positive, but less than 1?

Q13 How do you think the derivative is affected by the transformation?

Now you'll use the locus of point S to see how the derivative is affected.

Make sure that your h is still very small!

15. Select points S and R and choose **Locus** from the Construct menu. Make the locus a new color.

Here, you can hide f, g, and the secant line if your sketch is too cluttered.

Q14 *Show* f' and compare the plot of f' and the approximation of g' (the locus). How do your approximations of $g'(x)$ and $f'(x)$ compare?

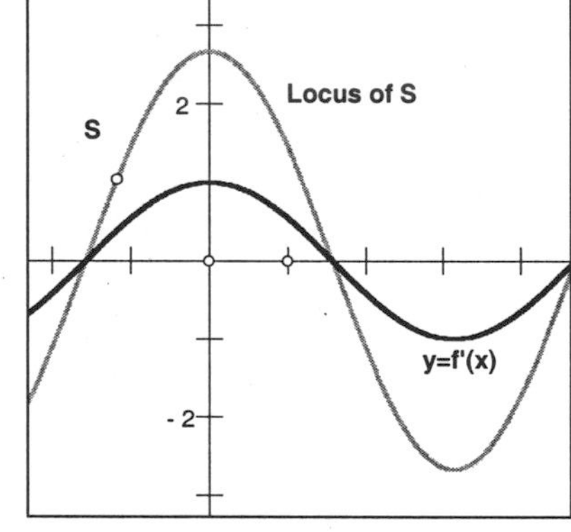

Q15 Write your prediction for $g'(x)$ in terms of $f'(x)$.

16. To check your answer, plot your prediction, then press the *Show* g' button.

If you don't see a new function when you press the *Show* g' button, then that means the two functions are identical and you are right!

Q16 Is the function you plotted in step 16 identical to the plot of $g'(x)$? If so, great! If not, see Q10 for other functions to try.

The last transformation possible is to multiply the x-value by a constant, but the procedure for finding out how this transforms the derivative is the same.

17. Go to page 4 and do steps 15–16 and answer Q11–Q16 for $g(x) = f(bx)$.

Q17 If you put all the transformations together, you'll get a function like $f(x) = 2\sin(3x) + 4$. Look at your answers for the various transformations and predict what the derivative of this function should be if all the rules you discovered above still work.

Q18 Now go to page 5 and press the *Transform* button to create a plot of the function $f(x) = 2\sin(3x) + 4$. Use your answer from above to write an expression for the derivative of this function. Then check to see if your prediction is correct.

Explore More

In this section, you'll use tangent lines to figure out how a transformation of $f(x)$ transforms its derivative. When you are using tangent lines to compare the different transformations and their derivatives, the goal will be to get equal slopes—in other words, parallel lines, if possible.

This is a duplicate of page 2 but without the constructions you made there.

1. Go to page 6. Press the *c→0* button and then the *Show Tangents* button.

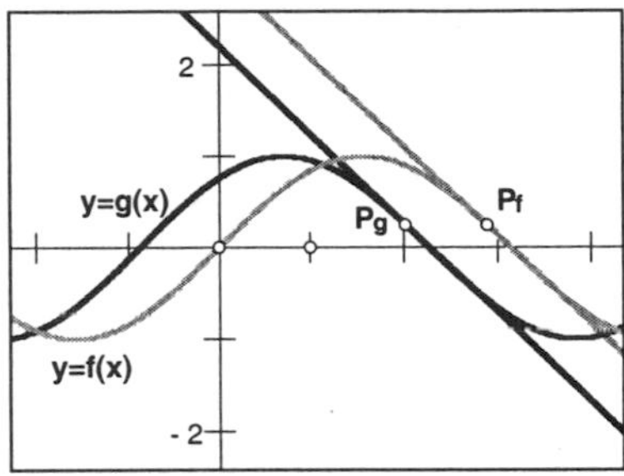

Q1 Adjust the slider for c. How does the tangent line to function $g(x)$ at point P_g move as c changes? How does the slope change?

Q2 What are the coordinates of point P_g in terms of the coordinates of point $P_f\,(x_P, f(x_P))$ and c?

Q3 Using your answers above, what is the equation of the tangent line to function $g(x)$ at point P_g? Check by plotting your new function.

2. Go to page 7. Press the *a→1* button and then the *Show Tangents* button.

Q4 Adjust the slider for a. How does the tangent line to function $g(x)$ at point P_g move as a changes? How does the slope change?

Q5 What are the coordinates of point P_g in terms of the coordinates of point $P_f\,(x_P, f(x_P))$ and a?

3. To figure out the exact slope of the tangent line, choose **Measure | Calculate** and make the new measurement $k \cdot slope_f$.

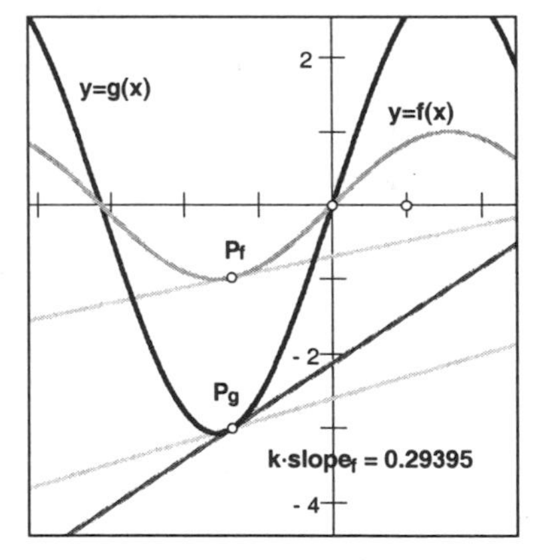

4. Then choose **PtSlopeLine** from **Custom** tools. Click on point P_g and then your new measurement.

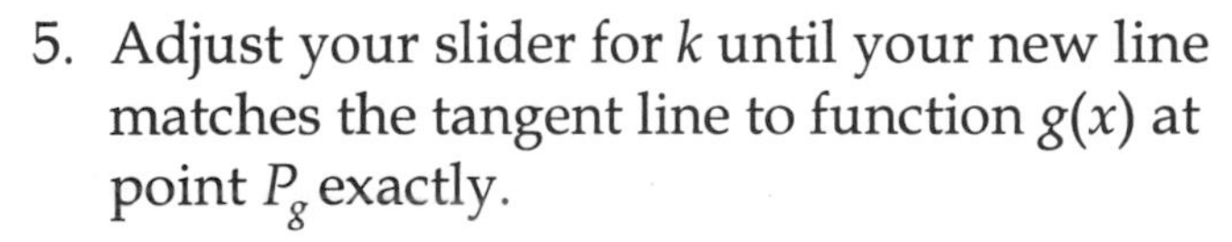

5. Adjust your slider for k until your new line matches the tangent line to function $g(x)$ at point P_g exactly.

Q6 What value did you get for k in order to match the two lines exactly?

Try the same procedure on page 8 to find the k for $f(bx)$.

Second Derivatives

Name(s): ______________________

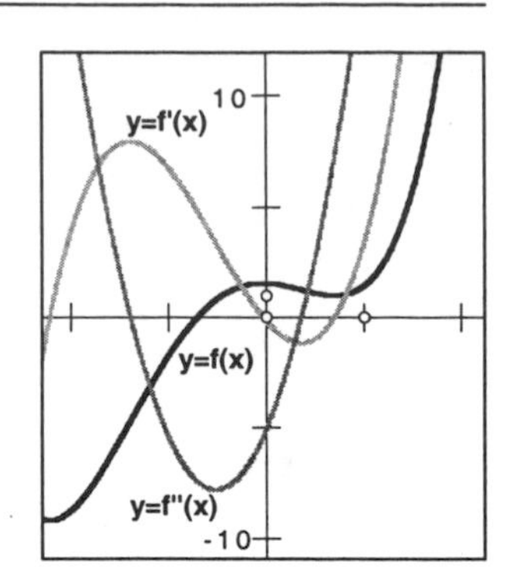

The derivative, $f'(x)$, of a function, $f(x)$, has been a very useful tool. It's given us much information about many things—the behavior of the function $f(x)$, the instantaneous rates of change at points on $f(x)$, and slopes of the function at a point. Since it has been so handy, why not do the process again—take the derivative of the derivative, or the *second derivative* of f. Does it give useful information about the original function as well as information about the first derivative? You will explore this question here.

Sketch and Investigate

You can edit the function f yourself at any time.

1. **Open** the document **SecondDerivative.gsp** in the **Exploring Derivatives** folder. On page 1 is a function $f(x)$, its plot, and various sliders. The sliders allow you to change the function's parameters.

2. You have learned the relationships between a function and its derivative. Press the *Show f′* button to refresh your memory. Then press the *Hide f* and *Hide f′* button and then the *Show f″* button.

You can also adjust the measurement for c with the appropriate slider.

3. Make the second derivative a constant function by pressing the *a→0* and *b→0* buttons.

Q1 The function f'' is the first derivative of f', so what should the plot of f' look like if f'' is a constant? Check your answer by pressing the *Show f′* button. Were you right?

Q2 Adjust the slider for c so that $c < 0$, then adjust the slider so that $c > 0$. Describe how this affects the first derivative $f'(x)$.

4. Press the *Hide f′* and then the *Show f* buttons. Adjust the slider for c so that $c < 0$, then adjust the slider so that $c > 0$.

Q3 Describe how adjusting the slider for c affects the function $f(x)$.

Now let's look at what happens if $f''(x)$ is a linear function.

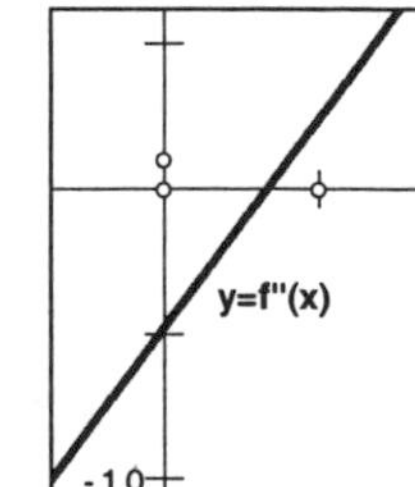

5. Make the second derivative an *increasing* linear function by adjusting the slider for b.

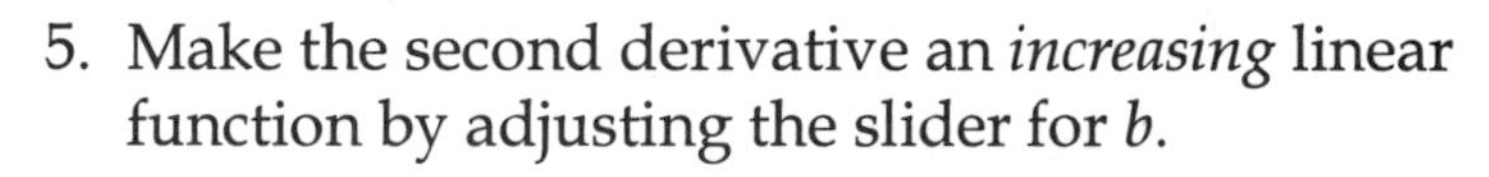

6. Press the *Show f′* button and then adjust the slider for b to make the second derivative a decreasing linear function.

Q4 Describe how adjusting the slider for b affects the first derivative.

Q5 When $f''(x) = 0$, or crosses the x-axis, what can you say about the plot of the first derivative? (Check both cases—$f''(x)$ increasing and decreasing.)

Q6 What is the significance of the root of $f''(x)$ for the function $f(x)$? Check your answer by trying different values of c.

Do the qualities you discovered hold for all functions? Let's look at one more case, where $f''(x)$ is a quadratic.

7. Hide $f(x)$ and $f'(x)$. Use the sliders for a, b, and c to create a new plot of $f''(x)$ where the function is always positive, but not constant.

Q7 Predict what you think will be true about $f(x)$ and $f'(x)$ in this case. Check your answers by showing f and f'. How did you do?

8. Hide $f(x)$and $f'(x)$ again. This time, use the sliders for a, b, and c to create a new plot of $f''(x)$ where the function is always negative.

Q8 Predict what you think will be true about $f(x)$ and $f'(x)$ in this case. Check your answers by showing f and f'. How did you do?

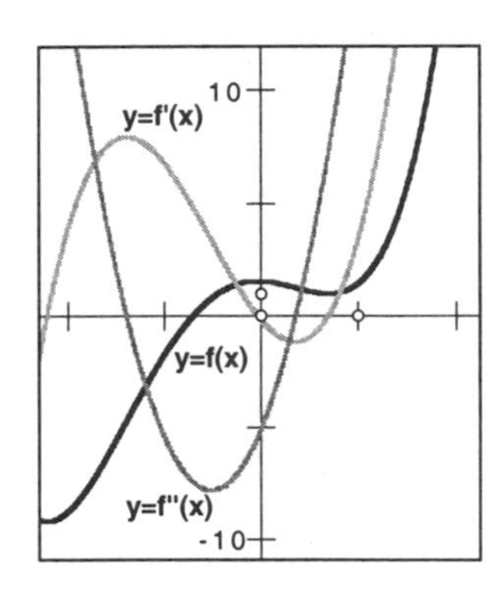

Q9 Make a chart of how $f''(x)$ affects both $f(x)$ and $f'(x)$. Where $f''(x)$ is negative, what happens to $f'(x)$ and $f(x)$? Where $f''(x)$ is positive, what happens to $f'(x)$ and $f(x)$? Where $f''(x)$ is 0, what happens to $f'(x)$ and $f(x)$?

It is easier to check your conclusions if you look at just two functions at a time.

9. Check your conclusions in Q9 by creating a new plot of $f''(x)$ where the function is sometimes negative and sometimes positive. How did you do?

On page 2 of the document, all three plots have the same color. Using what you learned above, can you identify which is which without using the buttons?

10. Predict which plot is $f''(x)$, $f(x)$, and $f'(x)$. Check your answer by pressing the *Show* buttons.

11. Press all three *Hide* buttons to hide the answers. Then, press the *Randomize Sliders* button to change the values of all the sliders. Press the button again to stop. Identify each of the plots again.

There is another relationship between the three functions to see. This one has to do with the tangent line to $f(x)$ at a point. You have seen that the tangent line's slope is the derivative at the point of tangency. But its position—above or below the function—also tells something important.

12. Go back to page 1. Use the *Hide/Show* buttons to make sure only $f(x)$ is showing.

13. Choose **Tangent Line** from **Custom** tools. Click anywhere on the plot of $f(x)$ and then on the expression for $f(x)$ to create a tangent line.

Q10 Show $f''(x)$ and then move your new point along the plot. When $f''(x)$ is negative, what can you say about the tangent line?

Q11 What happens to the tangent line where $f''(x)$ is positive? What happens at the points where $f''(x)$ changes from positive to negative?

Exploration 1

You can also use the limit of the secant line's slope.

In this section, you'll build the second derivative's plot using the limit of the difference quotient just like you built the first derivative's plot—the only difference is that $f'(x)$ is used in place of $f(x)$.

1. Go to page 3. On this page there are two extra sliders that adjust the measurements h_1 and h_2.

Enter any subscripts in brackets. So for D_1 you would enter D[1].

2. Choose **Graph | New Function** and enter the difference quotient for the first derivative: $(f(x + h_1) - f(x))/h_1$. Label this function D_1.

You can check your approximation by plotting your new function and then pressing the *Show f′* button.

3. Adjust the slider for h_1 so that the function $D_1(x)$ is a good approximation for $f'(x)$.

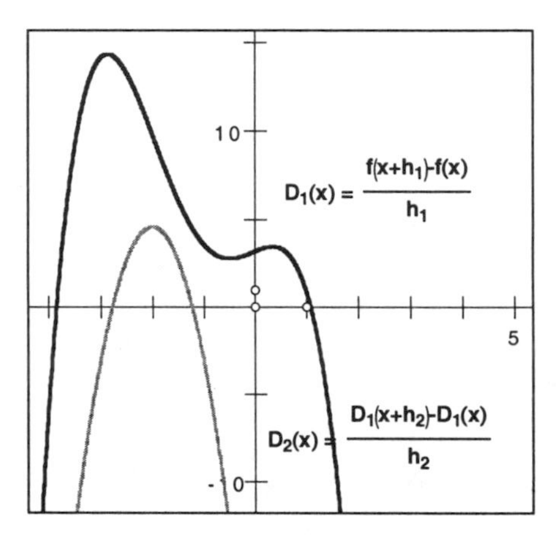

Now if you build the difference quotient using $D_1(x)$, this new function can be used to approximate the second derivative.

4. Choose **Plot New Function** from the Graph menu and enter the expression for the approximation of the second derivative: $(D_1(x + h_2) - D_1(x))/h_2$. Label this function D_2.

5. Press the *Show f″* button and then adjust the slider for h_2 so that the function $D_2(x)$ is a good graphical approximation for $f''(x)$.

6. To see if $D_2(x)$ is a good approximation numerically, choose the **Point** tool, create a point on the x-axis, and label it P. Measure the x-coordinate of point P, then choose **Calculate** from the Measure menu and enter the expression for the error: $f''(x_P) - D_2(x_P)$.

Q1 Were you able to make h_2 small enough so that the error was at most 0.01? (If not, drag the unit point on the x-axis to zoom in a bit).

Now that you have both difference quotients built, you can adjust any of the sliders a, b, c, or d to change the function $f(x)$ or you can double-click on $f(x)$ and enter any other function you'd like to try and the difference quotients will change dynamically as well.

Exploration 2

Go to page 4. This page has the plot of the second derivative of a function f, and its equation, $f''(x) = d + a\sin(bx + c)$. When you open the page, $b = 1$, and c, d, e, and $f = 0$. Show the plots of $f(x)$ and $f'(x)$. Probably no surprises there. Hide $f(x)$ and $f'(x)$. Adjust the slider for d so that $d \neq 0$ and predict what you'll see when you look at $f(x)$ and $f'(x)$. Show $f(x)$ and $f'(x)$. Surprised? What is the surprise and why is this so? (You may want to adjust d while $f(x)$ and $f'(x)$ are showing to answer this.)

Try this as well: On page 5, plot the functions $q(x) = x^2$ and $r(x) = \cos(x)$. What do you think the function $q(x) + r(x)$ looks like? Plot the combined function and see what happens. How does the second derivative explain the resulting graph? (Use the **Derivative** command twice to find the first and second derivatives of $q(x) + r(x)$ to visualize the answer.)

Exploration 3

On page 6 of the document, there is a sketch much like the one in "Visualizing Change: Velocity." On this page, you are able to control the acceleration of a point—the second derivative of position. If you set the acceleration to 0 by pressing the *a→0* button, then the sketch works just like the one in that activity. (You would use the velocity slider v_c.)

See "Visualizing Change: Velocity" for more information.

Here you can control the acceleration by adjusting the a slider. To try this out, set your initial velocity with the v_c slider. Then press the *Start Motion* button and adjust only the a slider. The other velocity slider, labeled *velocity*, will change with the acceleration.

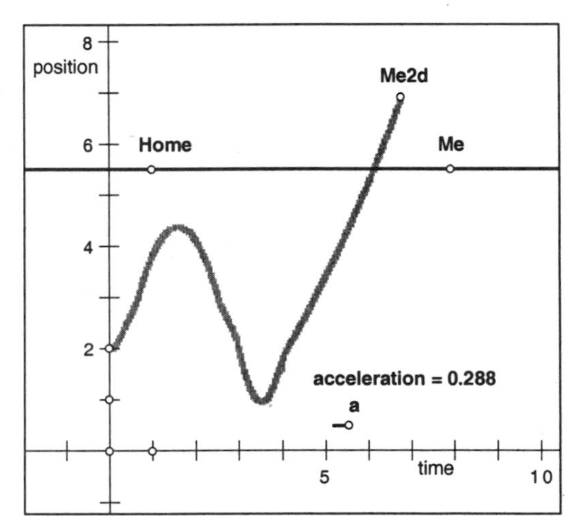

What happens to the velocity if you maintain a positive acceleration? What if you maintain a negative acceleration? Can point *Me* move in the positive direction when the acceleration is negative?

Try various actions with either slider to see how the position trace moves. Then you can try to match Path 1 or Path 2.

Newton's Method

Name(s): ______________________

While you can solve many equations with familiar algebraic techniques, there are also quite a few that you can't solve algebraically: $2^x = x^2$, for instance. From the plot, it looks like there are three solutions to this equation. The ones in the first quadrant you could get by guessing and checking, but probably not the one in the second quadrant. In this activity, you will learn an algorithm for solving equations using derivatives.

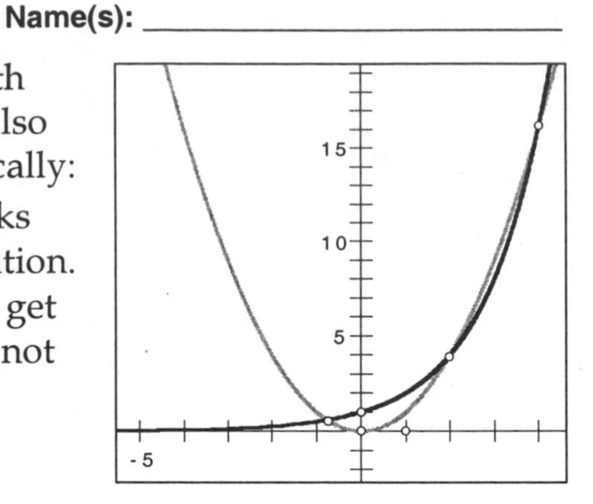

Sketch and Investigate

1. **Open** the document **Newton.gsp** in the **Exploring Derivatives** folder. On page 1 you will find the function $f(x) = 2^x - x^2$.

Q1 Why is solving the equation $2^x = x^2$ equivalent to finding the zeroes for $f(x) = 2^x - x^2$?

You have seen that a tangent line to a function at a point can be used to approximate a differentiable function on a small interval around that point. Going further with this idea, you can use the x-intercept of a tangent line to approximate the x-intercept of a function. First we'll see how this works and then you'll make a tool to do the process.

To start the process, we choose a starting point on the x-axis, called a *seed,* and find the point on the function plot that corresponds to that x-value.

To measure the x-coordinate, choose **Measure | Abscissa (x)**.

2. With the **Point** tool, construct a point on the x-axis close, but not too close, to the zero of $f(x)$ on the left side of the origin. Measure the new point's x-coordinate.

3. Calculate $f(x_A)$ by choosing **Calculate** from the Measure menu and click on the expression for f and then on the measurement x_A.

4. Plot the point $(x_A, f(x_A))$ by selecting measurement x_A, then $f(x_A)$, and choosing **Plot As (x, y)** from the Graph menu.

The next step is to construct a tangent line to $f(x)$ at the point $(x_A, f(x_A))$.

You can also click on the point $(x_A, f(x_A))$ with this tool.

5. Choose **Tangent Line** from **Custom** tools and click on point A and then the expression for $f(x)$.

We now need to find where this tangent intersects the x-axis. This intersection point, or *root* of the tangent line, represents the first approximation for the zero of the function.

6. With the **Arrow** tool, construct the intersection of the tangent line with the x-axis by clicking on that spot. Label this point B.

Select point B with the **Arrow** tool and choose **Abscissa (x)** from the Measure menu.

7. Measure the x-coordinate of point B.

Our goal here is to approximate the zero of the function as closely as possible, so we want to repeat this process until the tangent line's root, point B, is indistinguishable from the function's root.

8. Select point A and point $(x_A, f(x_A))$. Then choose **Segment** from the Construct menu. Go to the Display menu. Make the segment a different color from the Color submenu, and choose **Thick** from the Line Width submenu.

To hide the line, select it and choose **Hide Line** from the Display menu.

9. Hide the tangent line and construct a line segment between the point $(x_A, f(x_A))$ and point B.

10. Select point A and then point B. Choose **Edit | Action Buttons | Movement.** On the Move panel, set speed to **instant**.

11. Select both line segments, then turn tracing on by choosing **Display | Trace Segments.**

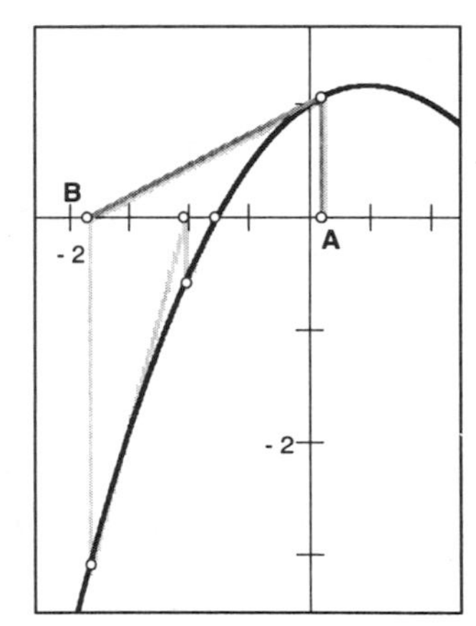

12. Press the *Move A→B* button, and you should see the second iteration. Using the *Move* button, continue "zeroing in" on the root until the x-coordinate of point B is constant to the hundred-thousandths place.

Q2 What final value did you find for the measurement x_B?

Hint: If x_B is a root, what is the value of $f(x_B)$ supposed to be?

Q3 Check your answer by calculating $f(x_B)$. What was your approximation error?

The disadvantage of using a *Move* button and a trace is that if you move your original points, the traces do not move with them. So it is time for a tool.

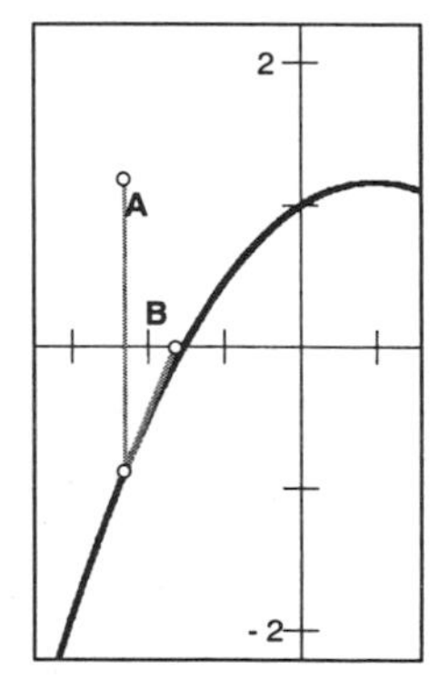

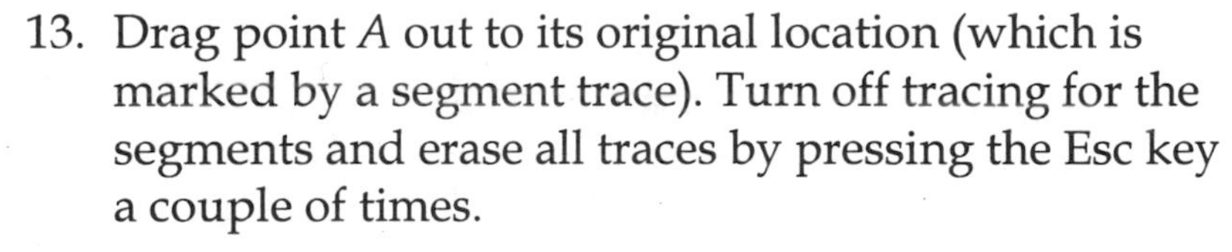

13. Drag point A out to its original location (which is marked by a segment trace). Turn off tracing for the segments and erase all traces by pressing the Esc key a couple of times.

14. With just point A selected, choose **Split Point From Axis** from the Edit menu.

15. To create the tool, select the givens: point A, the expression for $f(x)$, and the x-axis. Select the results: point B, point $(x_A, f(x_A))$, both segments, and the measurement x_B. Then choose **Create New Tool**

from **Custom** tools. Name it **Newton** and check Show Script View. Click OK.

16. Click in an empty spot of the sketch to deselect all objects. Select point A and the x-axis, and choose **Edit | Merge Point To Axis.**

17. In the tool's Script View, double-click given Function f and check Automatically Match Sketch Object. Click OK. Do this again for given Straight Object x.

After you're done, Function f and Straight Object x will be assumed as shown.

Assuming:
1. Marked Coordinate System S_1
2. Function f
3. Straight Object x

Now you are ready to use the new tool. If point B is too close to the root or point A to be distinct, use the **Arrow** tool to move point A until point B is distinct.

18. Choose **Newton** from **Custom** tools and click on point B. The tool will construct a second iteration.

19. Using the tool again, click on the intersection point of the new tangent segment with the x-axis to construct a third iteration.

20. Continue using the tool until you have at least five iterations. If at any stage an intersection is not distinct, adjust point A's location before using the tool again.

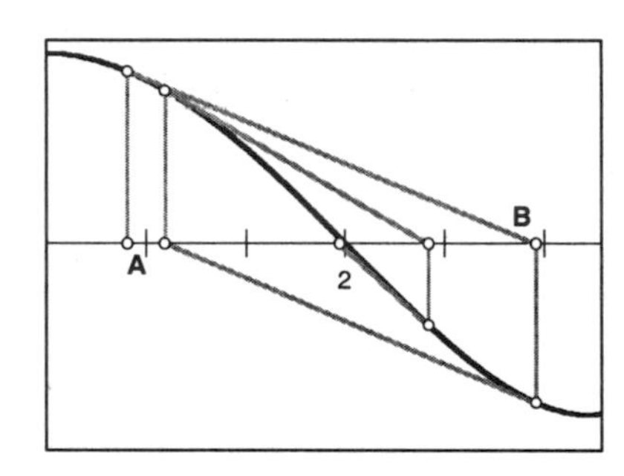

Q4 Adjust point A (the seed) so that it's fairly close to a root of the function. What can you say about the convergence of Newton's Method when the seed is close to a root?

Q5 Move the seed value so it's to the left of the leftmost root or to the right of the rightmost root. What can you say about the convergence of Newton's Method in these situations?

Q6 Try some different positions for the seed value between the various roots. What can you say about the stability and convergence of the method under these conditions?

21. Go to page 2, where there is a fourth-degree polynomial. Repeat Q4–Q6 with this function and see if your previous answers hold up.

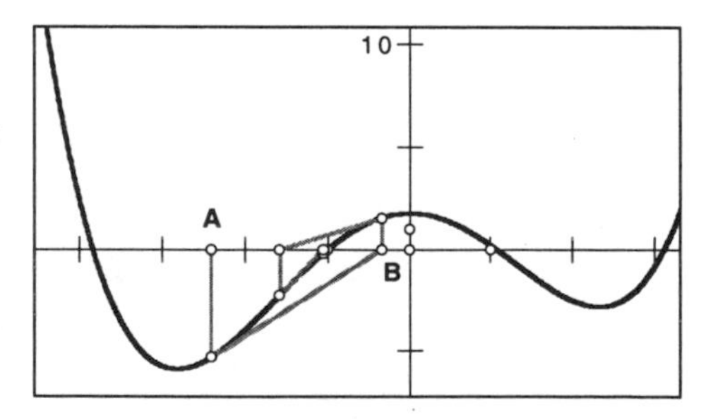

Q7 Adjust your sliders so that the function has only a single root, but also has a minimum or maximum that comes close to the x-axis without touching it. Notice what happens to the iterations as you move the seed value to different positions near this minimum or maximum. What can you conclude about the stability of Newton's Method in this situation?

Q8 Can you think of a way to predict how well Newton's Method will do with different seed values, depending on the shape of the graph? Describe your predictions as clearly as you can.

Explore More

On page 3 of the document, another function is plotted, here with the first step of Newton's Method done. This time you'll use Sketchpad's **Iterate** command to create the successive approximations.

1. Select point *A* and choose **Iterate** from the Transform menu. Click on point *B*. Uncheck Tabulate Iterated Values in the Structure pop-up menu. Press the plus (+) key twice to increase the number of iterations to 5. Click **Iterate**.

A good point to click on is the point $(x_B, f(x_B))$.

Select the iterated image and press the plus (+) key to increase the number of iterations.

2. Select the iterated image of point *A* so that only the points are selected. Choose **Terminal Point** from the Transform menu. Measure the coordinates of this point. This is the value found by Newton's Method after the number of iterations you have set. You can increase the number of iterations at any time, and this point will remain at the last iteration.

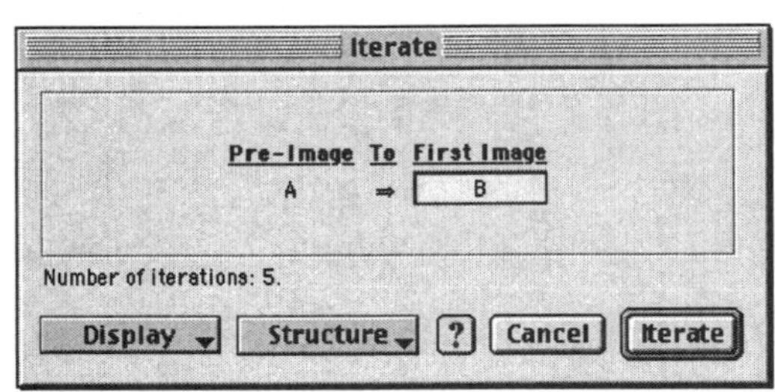

Q1 Continue your explorations on pages 4, 5, and 6. You only need to move point *A*. In what situations does the algorithm not bring you to an *x*-intercept? Explain in your own words why this happens.

Q2 If point *A*'s *x*-coordinate is labeled x_0, what is the equation for the first tangent line you constructed?

Q3 If point *B*'s *x*-coordinate is labeled x_1, solve your equation in Q2 to find a formula for finding x_1 in terms of x_0 and the function $f(x)$.

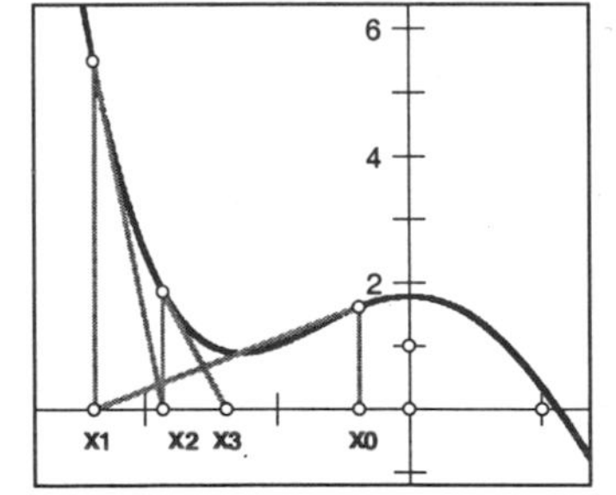

Q4 For the next iteration, the algorithm starts over and now uses x_1, so write a formula for finding x_2 in terms of x_1 and $f(x)$.

Q5 Generalize your work above and write a formula for finding x_{n+1} in terms of x_n and $f(x)$.

4

Exploring Antiderivatives

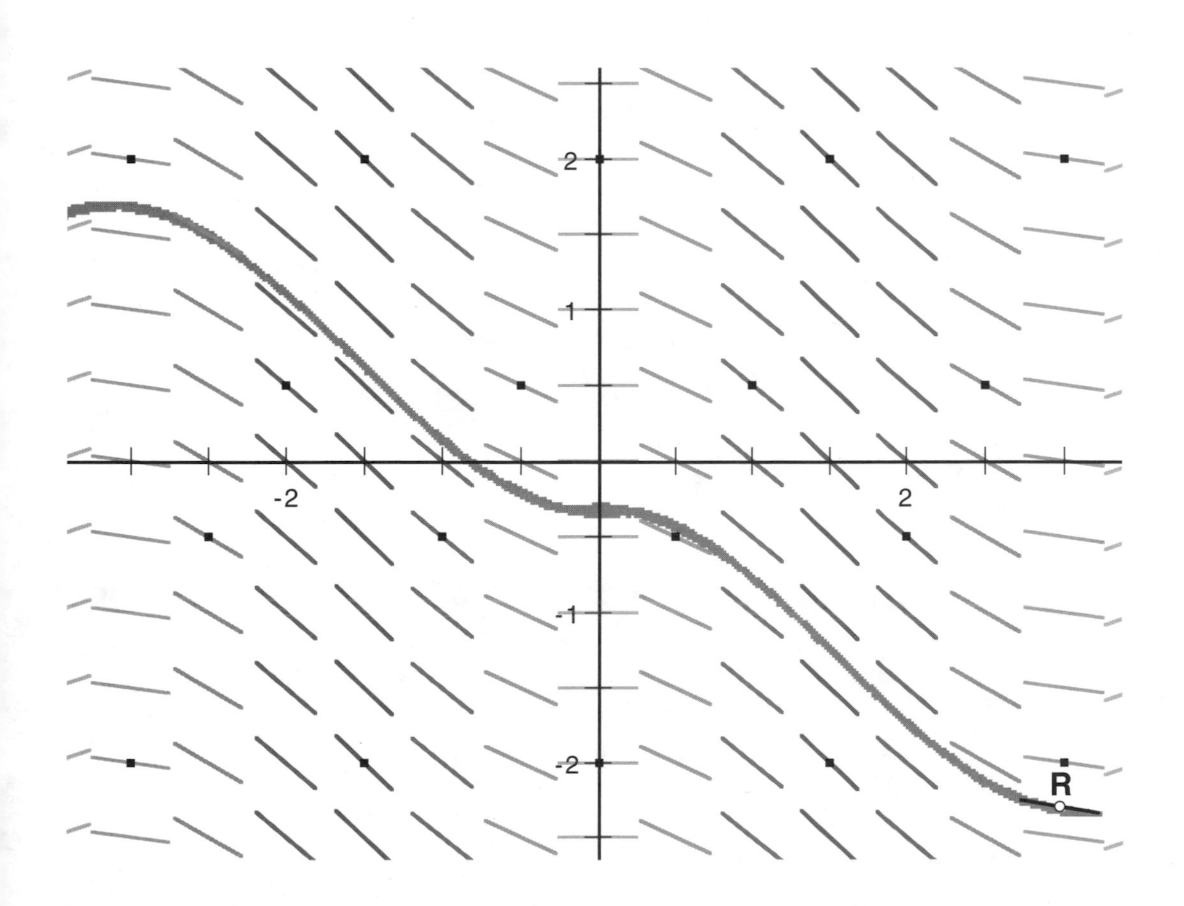

Following Tangent Lines

Name(s): ____________________

In previous activities, one of your goals was to plot the derivative of a given function f and examine how the two functions f and f' were related. Here, you will again start with a plot, but this time your plot will be the derivative, or slope function, of some unknown function. Your job is to determine the shape of the unknown function when given its derivative. Because you're starting with the derivative and working backward, you're trying to find the *antiderivative* of the function you're given.

Sketch and Investigate

The best tools for predicting what a plot of $f(x)$ might look like when you have the plot of its slope are tangent lines. First, let's review what tangent lines do for us when we *do* have f.

1. **Open** the document **FollowTangent.gsp** in the **Exploring Antiderivatives** folder. You will find a function $f(x)$, its plot, a tangent line to $f(x)$ at point R, and a plot of the derivative of $f(x)$.
2. Move point R along the plot of $f(x)$ and watch how the little arrow at point Q helps show the shape of the plot—even though Q is not a point on the function plot. When you are done observing, move point R as far left as you can. Do you really need f if you have the little arrow at point R?
3. With point R selected, choose **Edit | Split Point From Function Plot** and press the *Hide* $f(x)$ button.
4. Select point R and choose **Trace Point** from the Display menu.

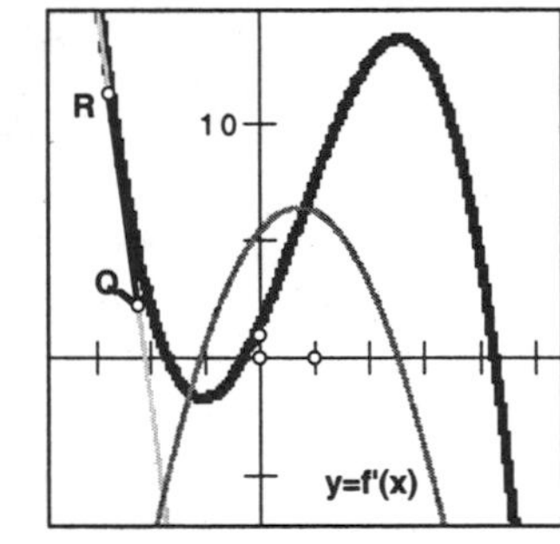

Remember, you already know what f looks like.

5. Can you retrace the now hidden plot of f by moving point R in the direction of point Q? Now that point R is not on the function plot, you'll need to move it carefully in the direction indicated by the arrow. Give it a try!

Q1 Press the *Show* $f(x)$ button. Could you trace out the basic shape of f? What might account for inaccuracies in your trace?

Q2 Write the equation of the tangent line through point R (x_R, y_R) in point-slope form, using derivative notation for the slope.

If you follow the tangent line precisely, you can create a very good rendering of the shape of the function f. However, you also want your prediction of the antiderivative to be fairly accurate. Just how accurate is using the tangent line to approximate values?

If you want to change the location of point R, edit the parameter x_R.

6. To review the answer to this question, go to page 2 of the document. Here you have the same function, tangent line, and point R. Point R is fixed this time as the "zoom-in center" at $x = 2$.

7. The error you get using the tangent line's y-coordinate instead of the function is represented by the red segment between point S and point S_T. Move point S so that $x_S \approx 1$.

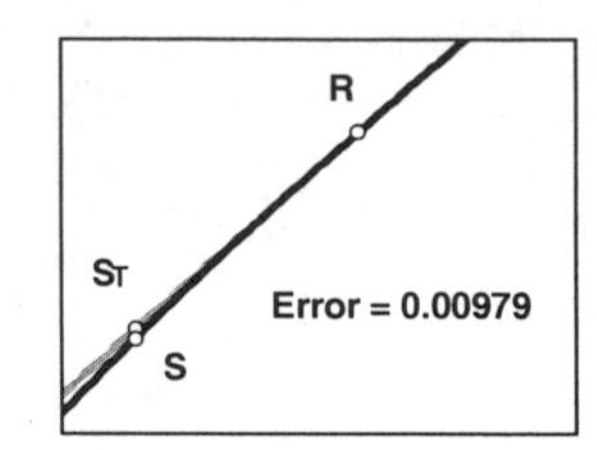

Q3 Use the x- and y-*scale* sliders to zoom in on this function at point R. Compare the y-coordinate that you'd get using the tangent line to the function's y-coordinate. How close do you need to get to $x = 2$ to make your error at most 0.01?

You can see that the tangent line to $f(x)$ at $x = x_R$ is a good approximation of $f(x)$ near R. How is this useful? In the next few steps, you will use the derivative to construct a tangent line to a function without the function being there and use it to sketch the plot of the function.

8. Go to page 3 of the document. Here you will find the plot of a function $f'(x)$ and point R.

Select point R and then choose **Abscissa (x)** from the Measure menu to measure the x-coordinate and **Ordinate (y)** to measure the y-coordinate.

9. You need to construct a tangent line at point R, so measure its x- and y-coordinates.

10. If you assume that point R is on $f(x)$, then its y-coordinate is $f(x_R)$. To relabel its y-coordinate as $f(x_R)$, double-click on measurement y_R with the **Text** tool and enter `f(x[R])` on the Label panel.

You do not have the equation or a plot of $f(x)$, but you can still plot its tangent line at point R because you have its slope.

11. Where is this slope? You can get it from the given function $f'(x)$. Choose **Measure | Calculate** and calculate the slope of the tangent line $f'(x_R)$. (What does $f'(x_R)$ represent on the plot of $f'(x)$?)

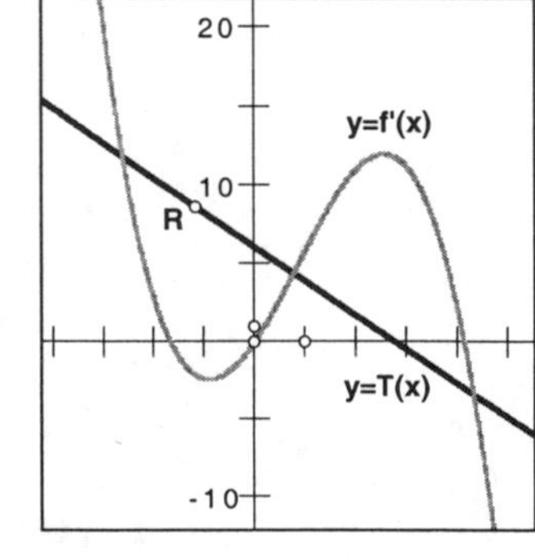

12. Choose **Graph | Plot New Function** and enter the expression of the tangent line in point-intercept form: $f(x_R) + f'(x_R) \cdot (x - x_R)$. Label this function T.

You also have the value of $f'(x_R)$ so that you can observe the exact value of the slope of the line as you move point R.

13. Move point R around the window and observe how the slope of the tangent line changes.

Q4 Where does this line have a negative slope? Where does it have a positive slope? Where does it have a slope of 0?

You can move a point vertically using the Up Arrow key.

Q5 Move point R vertically only. What do you observe about the slope of the line? Why does this happen?

You have created a line, $T(x)$, whose slope is equal to $f'(x_R)$. While this line has the equation of the tangent line for our unknown function $f(x)$ at point R, it is not actually "tangent" to anything at the moment.

Let's assume for the moment that point R is on our function f. Then, for points near $x = x_R$, this line gives a very close approximation of the plot of $f(x)$. Can you use this line to sketch the plot of the function $f(x)$? Yes! But only if you can follow the tangent line for a short distance. In order to make a small segment on the tangent line, you will use the slider that adjusts the measurement h.

14. Choose **Calculate** from the Measure menu and create the expression $x_R + h$.

15. To plot a second, nearby point on the tangent line, start by calculating the value of $T(x_R + h)$.

Click once in an empty spot to deselect an object.

16. Deselect the measurement $T(x_R + h)$ and then select the measurements $x_R + h$ and $T(x_R + h)$ in that order and choose **Graph | Plot as (x, y).** Label this point Q.

17. With point Q selected, select point R and choose **Segment** from the Construct menu.

18. With the segment selected, turn on tracing by choosing **Trace Segment** from the Display menu.

Now the segment will leave a trace behind as you move point R around the plane. This segment will trace out the shape of the original unknown function (the antiderivative), if you always move point R in the direction of point Q.

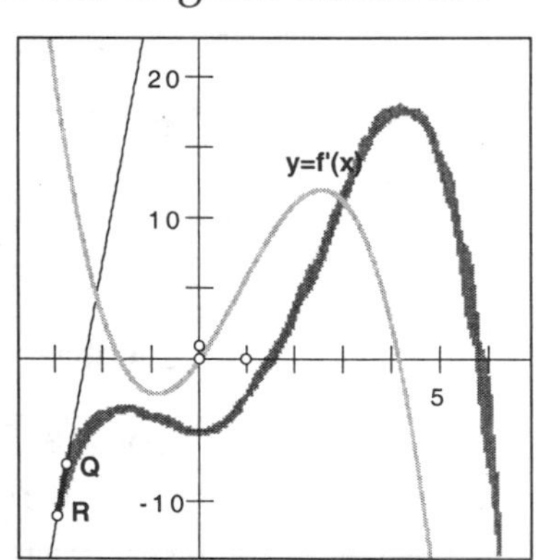

If you start out too high or too low on the screen, your trace will run off the edge of the screen. If so, erase the traces and start again in a different position.

19. Select point R and drag it slowly in the direction of point Q. At the slightest movement of point R, however, the slope of the line will change, so you must constantly adjust, changing the direction in which you move point R as you go. (This may take some practice!)

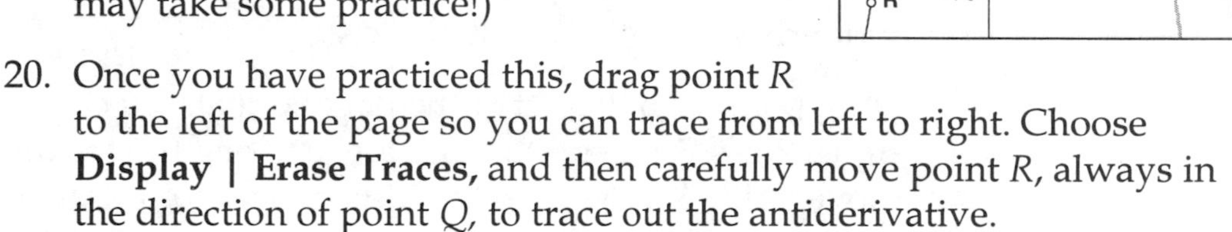

20. Once you have practiced this, drag point R to the left of the page so you can trace from left to right. Choose **Display | Erase Traces,** and then carefully move point R, always in the direction of point Q, to trace out the antiderivative.

You may have trouble at first tracing in the correct direction. Keep trying, and you'll get better at following the direction of point Q.

Q6 Examine the trace you have created. Where is the function it represents increasing? Where is it decreasing? Where does it reach a local maximum or minimum?

Q7 How do your answers to Q6 compare to your answers to Q4?

The accuracy also depends on h being relatively close to 0. The closer to 0, the more accurate the trace.

The accuracy of the antiderivative you've traced so far depends on your ability to follow the direction of point Q. You can automate this process by having point R try to move toward point Q. This will guarantee that it goes in the correct direction. What's more, point R can never catch up with point Q, so it will keep moving toward it indefinitely.

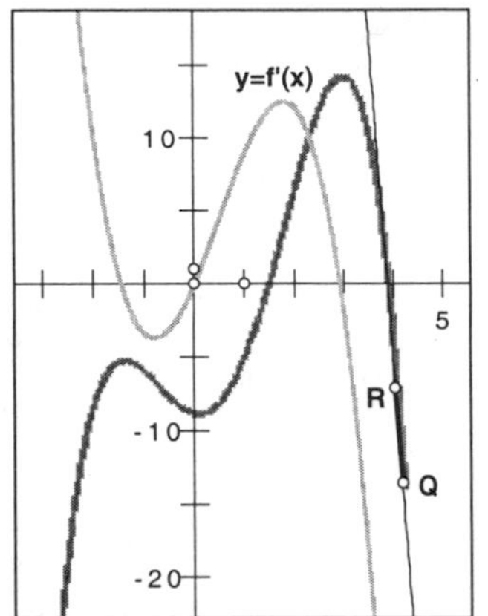

21. Erase your traces. Select point R and point Q, in that order. Choose **Movement** from the Action Buttons submenu of the Edit menu and keep the speed at **medium**. Click OK.

If you start out too high or too low on the screen, your trace will run off the edge of the screen. If so, erase the traces and start again in a different position.

22. Position point R where you want tracing to start, erase traces and then press the *Move Point* button and watch it go.

How accurate was your trace compared with the "automatic" trace?

23. Choose **Undo Animate Point** from the Edit menu, and point R will return to its initial location.

24. Make several traces, starting from the same horizontal position but from different vertical positions. To do this, turn off tracing for the segment and move point R up or down vertically, then turn on tracing again.

Q8 What do you notice about the traces? Can all these traces be approximations for the antiderivative? If so, how can this be? If not, explain why they are not.

Explore More

Look at the functions on pages 4–8. Each page has a different kind of function as the given slope function. For each one, make a prediction for the antiderivative, then follow these steps:

1. Drag point R to the left of the page so that you can trace from left to right. Press Esc twice to erase any existing traces, and then carefully move point R, always in the direction of point Q, in order to trace out the antiderivative. Repeat this step until you have a relatively smooth trace that you like.

2. Choose **Edit | Undo Animate Point** to return point R to your initial starting point. Then press the *Move R→Q* button. Compare your trace with the automatic trace.

Step to the Antiderivative

Name(s): ______________________

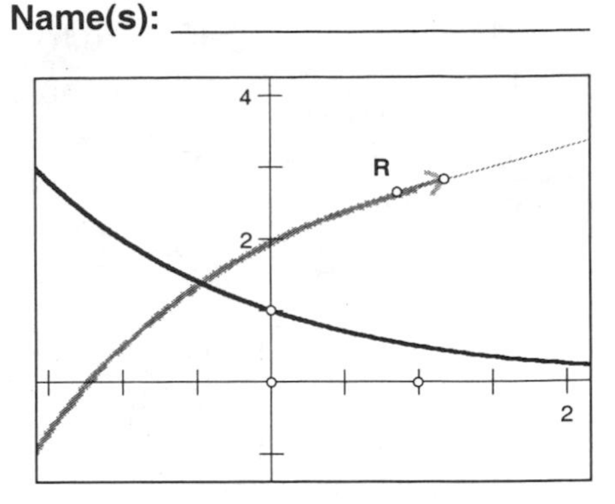

The fact that the tangent line to a function at a given point is a close approximation of the function near that point is a very useful thing. In the previous activity, you used tangent lines to determine the shape of the plot of $f(x)$ given a plot of its derivative, $f'(x)$. In this activity, you'll use tangent lines to construct tools that will create dynamic approximations of the antiderivative, rather than traces. In the process, you'll look rather closely at the construction of the approximate antiderivative so that you can train yourself in plotting antiderivatives by hand.

Sketch and Investigate

1. **Open** the document **StepTangent.gsp** in the **Exploring Antiderivatives** folder. On page 1 you will find a function plot $f(x)$, a few sliders, and an independent point R. Suppose that the function $f(x)$ is the derivative of some unknown function F, so $F'(x) = f(x)$.

If you did not do the last activity, go to page 89 for more in-depth explanations.

2. You will also find line $T(x)$ through point R. This line has a slope equal to $F'(x_R)$, or $f(x_R)$—so $T(x) = y_R + f(x_R)(x - x_R)$—and point Q with coordinates $(x_R + h, T(x_R + h))$, just as you built in the last activity.

Q1 Verify that the line $T(x)$ does have slope $F'(x_R)$, or $f(x_R)$, using the appropriate measurements from the plots and point R. Check that your verification holds even if you move point R.

To trace the location of point R as it follows point Q, you can press the *Move R→Q* button.

For points near $x = x_R$, this line gives a very close approximation of the plot of $F(x)$, and in the previous activity, you approximated $F(x)$ by tracing point R as it followed point Q. This time, you'll construct a tool that creates a tiny segment of that line so you can build an approximate plot of $F(x)$, one segment at a time, simulating how you might sketch an antiderivative with paper and pencil.

3. Construct a segment from point R to point Q by selecting point R and point Q and choosing **Segment** from the Construct menu.

To hide point Q's label, click once on point Q with the **Text** tool.

4. Select line $T(x)$ and choose **Display | Hide Line.** Hide point Q's label.

5. To create the new tool, select the givens: point R, measurement h, and the expression for $f(x)$. Then select the results: point Q and the segment RQ. Choose **Create New Tool** from **Custom** tools and name this tool **Antiderivative Segment.** Check Show Script View and click OK.

After you're done, Function f and Measurement h will be assumed as shown.

Assuming:
1. Marked Coordinate System S_1
2. Measurement h
3. Function f

6. In the tool's Script View, double-click given Measurement h and check Automatically Match Sketch Object. Click OK. Do this again for given Function f.

7. Choose **Antiderivative Segment** from **Custom** tools. Click on point Q.

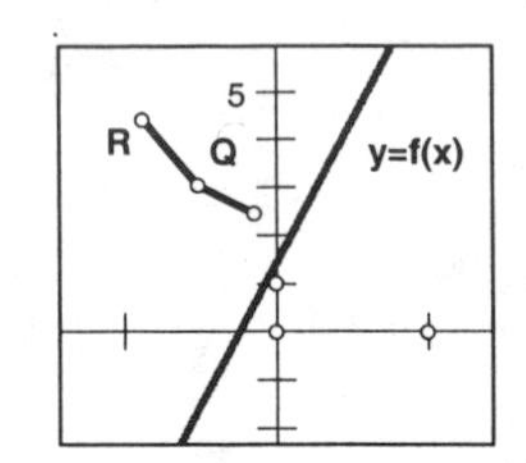

When you clicked on point Q with the tool, you made a new segment that starts at point Q and has a slope equal to $F'(x_Q)$, or $f(x_Q)$. By repeating this process, using each new point to make a new segment, you can build an approximation for $F(x)$ one step at a time.

Q2 Suppose you had been doing this with paper and pencil, starting with point Q, and you wanted to draw the segment constructed in step 7. What tells you the slope of your segment? How could you have predicted, in the figure above, that the second segment would have been angled as it is shown?

Q3 For your function, predict what the third segment will look like and draw a sketch of it in the margin. Then check your answer by using the tool to create the third segment.

8. Adjust the slider for h so that $h \approx 0.25$. Then create at least ten more segments by clicking on the endpoint of the previous segment with the **Antiderivative Segment** tool.

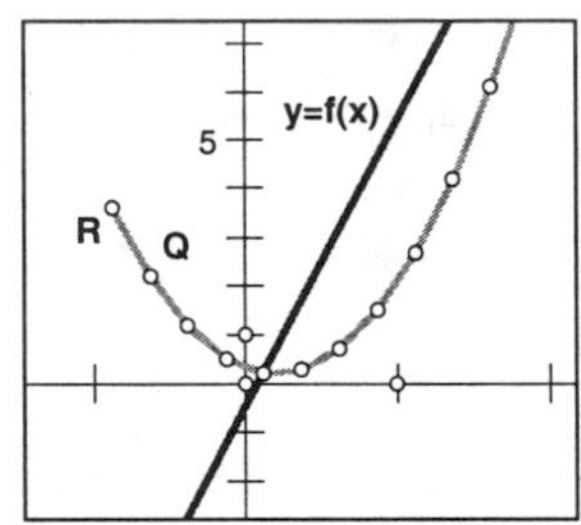

Q4 Suppose you had been doing this with paper and pencil. How could you have predicted when a segment would have a positive slope? A negative slope? A slope of 0?

Q5 How do you know when to make a segment whose slope is greater than the previous segment? Less than the previous segment?

9. Go to page 2 of the document, and either print the page or make a rough sketch of the function on your own paper. Starting from point A, sketch the antiderivative of the function shown, either smoothly or as a series of segments.

10. Adjust the slider so that $h \approx 0.25$. Choose the **Antiderivative Segment** tool and construct enough segments, starting at point A, to cover the interval $[x_A, 2]$. Compare this with the sketch you made by hand.

Point Q's label is not showing because point Q is very close to point R.

11. To compare your approximation with a very good approximation of the antiderivative, press the *Show RQ* button. Press the *Move R→A* button and then press the *Move R→Q* button.

Q6 How close was your hand-drawn sketch for the approximation of $F(x)$? How close was your approximation using the tool?

Q7 Adjust the slider for h so that $h \approx 0.1$. Describe what happens to the approximation you constructed with the tool and its accuracy. Explain why this happens.

You can also move point *A* up or down by pressing the Up or Down Arrow keys.

12. Adjust the slider for h so that $h \approx 0.25$ again. Select point A and drag point A up or down to a new location with the same x-coordinate. Press the *Move R→A* button and then the *Move R→Q* button to make a new trace at this new location.

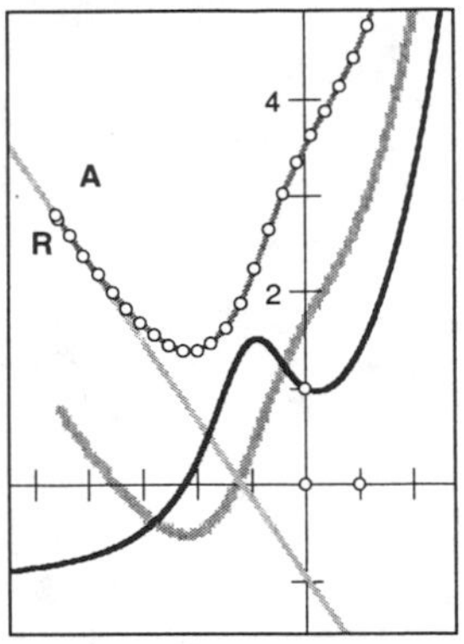

Q8 What is the relationship between the new trace and the previous trace? Use your knowledge of derivatives to explain this relationship.

You can also use point *R* and the *Move* buttons to make a trace.

13. On printed copies for each of pages 3–6, sketch by hand the antiderivatives of the given function plots. Then, go to the sketch and use the tool to construct an approximation.

Q9 On page 6, the y-scale is not shown. Why is it still possible to sketch an approximate plot of the antiderivative of this function?

Use your work from pages 2–6 to answer the following questions.

Q10 When $f(x)$ is positive, what can you say about your antiderivative approximation? What can you say when $f(x)$ is negative?

Q11 When $f(x)$ is increasing, what can you say about your antiderivative approximation? What can you say when $f(x)$ is decreasing?

Q12 Where $f(x)$ changes from positive to 0 to negative, what can you say about your antiderivative approximation? What can you say about where $f(x)$ changes from negative to 0 to positive?

Explore More

When you constructed the tool on page 1 of this document, you selected the expression for $f(x)$ as one of the givens. So, if all you have is the plot of $f(x)$, this tool won't work. We need a different tool for that case.

1. Go to page 7 of the document. On this page, segments and a semicircle define $f(x)$. Try using your tool here. What happens?

Q1 When you sketch an antiderivative by hand, you don't really need the expression of $f(x)$—just the y-value. Sketch a graph of an antiderivative of the function $f(x)$ shown on page 7.

Q2 What happens at the points where $x = -2, 2, 4,$ or 6?

In the steps below, you will make a new tool that addresses this case.

The point should be labeled *A* automatically. If not, use your **Text** tool to label this point *A*.

2. With the **Point** tool, construct a point on the semicircle. Measure the y-coordinate of this point by choosing **Measure | Ordinate (y).** Since

we lack an expression for $f(x)$, we can use this value, y_A, for the slope of the segment used in this tool.

3. Select the point again and measure its x-coordinate by choosing **Measure | Abscissa (x).**

4. With the **Point** tool, create an independent point, R, in the plane. Measure its x- and y-coordinates as you did above.

Compare this with the definition used in step 2 of Sketch and Investigate.

5. This time, define the tangent line function by choosing **Graph | New Function** and entering $y_R + y_A(x - x_R)$. Label this function $T(x)$.

In place of the measurement h used earlier.

6. On a small interval, this linear function $T(x)$ will approximate the plot of the antiderivative. To define the length of this interval, create a second point anywhere on the semicircle. Measure the x-coordinate of this point, labeled C, and calculate $x_C - x_A$. Label this calculation h.

7. Choose **Measure | Calculate** and create the expression $x_R + h$. Then, calculate $T(x_R + h)$. Select $x_R + h$ and $T(x_R + h)$ and choose **Graph | Plot as (x, y)**. With the new point still selected, select point R and choose **Construct | Segment.**

8. Select points A, C, R, the new point constructed in step 7 (in that order), and the segment. Choose **Create New Tool** from **Custom** tools. Name it **Anti-Segment2.**

Your tool can now be used to construct an antiderivative of this and any other function with just the function's plot.

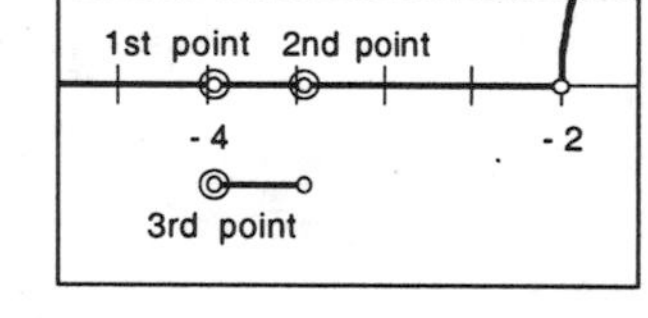

9. To use the tool, go to page 8. Choose **Anti-Segment2** from **Custom** tools. Construct two points on the first line segment of the function plot and then construct a third point directly below your first constructed point, as shown.

A segment with a slope equal to the y-value of the first point is created. The spacing between the two points on the function plot determines the segment's length.

10. Before you create more of the approximation of the antiderivative, make sure your first and third points have the same x-coordinate.

Q3 Why is it necessary that their x-coordinates are the same? Why did you not need to worry about this with the previous tool?

Make sure that your first two points are always on the function plot.

11. Use your tool to construct the antiderivative. So that your constructed antiderivative is all one piece, match your first and third points with the endpoints of the previous segment as shown above.

Q4 How close was your original sketch in Q1 to your constructed antiderivative?

Plotting the Antiderivative

Name(s): ____________________

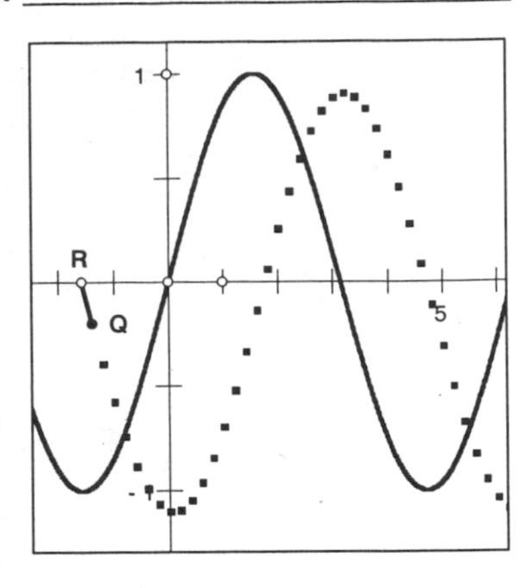

The process of constructing an approximate antiderivative, one linear step at a time, is called *Euler's method*. In the previous activity, it was very useful sketching antiderivatives, given the function's plot. In this activity, you will work with this process again—but this time with two different goals in mind. You will use Euler's method numerically to approximate the value of an antiderivative at any particular value of x and graphically to get an idea of the antiderivative expression.

Sketch and Investigate

1. **Open** the document **Antiderivatives.gsp** in the **Exploring Antiderivatives** folder. On page 1 of this document you will find a function $f(x)$, its plot, a fixed point R, and a parameter h. Here $f(x)$ is the derivative of some unknown function F, so $F'(x) = f(x)$.

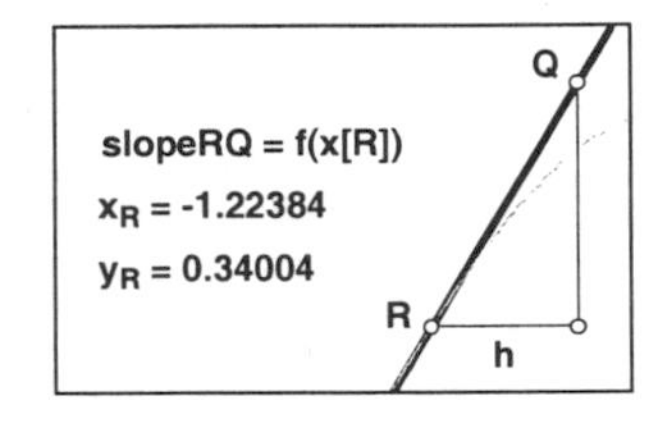

Q1 Suppose you are given initial values x_R and y_R of a point on $F(x)$. Using $f(x_R)$ as the slope of the plot of $F(x)$ at point R, what is an approximate value of $F(x)$ at $x_R + h$, shown in the figure as point Q?

Euler's method is an iterative algorithm: start with an x- and y-coordinate, perform the algorithm, and then repeat on the new coordinates. In Sketchpad, you can actually see this process using the **Iterate** command.

2. Here, point R has coordinates determined by the parameters x_R and y_R. Select the measurements x_R and y_R, and then choose **Iterate** from the Transform menu. Click on $x_R + h$ and then $y_R + f(x_R) \cdot h$ to match the pre-images to their first images. Uncheck **Tabulate Iterated Values** in the Structure pop-up menu. Press the *Iterate* button.

You can select the iterated image by clicking on any of the three image points with the **Arrow** tool.

3. The first three iterations of Euler's method will appear in your sketch. With the iterated image selected, choose **Properties** from the Edit menu. Change the number of iterations to 100 on the Iteration panel.

You can check your answer by plotting your prediction as a new function.

Q2 Using just the iterated plot given by Euler's method in step 3, can you predict what the antiderivative might be?

Next, we'll use Euler's method to go step by step to approximate $F(2)$ given that $f(x) = \sin(x)$ and we start at $(-2, 1)$.

This requires us to figure out how many steps there are of size h from the initial point to the terminal point. In this example, if the initial point has

$x_R = -2$, final point at $x = 2$, with step size 0.1, then $(2 - (-2))/0.1 = 40$, so we need 40 iterations.

4. Select the iterated image by clicking on any of the points in the image; then choose **Properties** from the Edit menu and change the number of iterations to 40 on the Iteration panel.

You can select an iterated image by clicking on any of the image points with the **Arrow** tool.

5. To construct the terminal point, choose **Terminal Point** from the Transform menu with the iterated image selected.

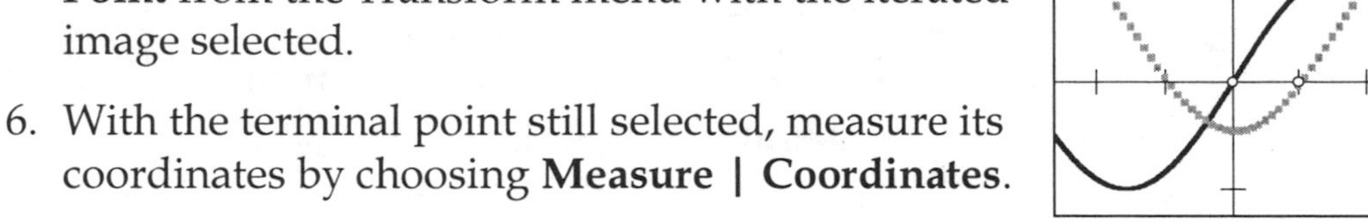

6. With the terminal point still selected, measure its coordinates by choosing **Measure | Coordinates**.

Q3 Is the terminal x-value after 40 steps of size 0.1 indeed 2? If so, what is the approximate value for $F(2)$ using Euler's method with a step size of 0.1?

To change a parameter's value, double-click on it with the **Arrow** tool and enter a new value.

Q4 What is the approximate value for $F(2)$ using Euler's method with a step size of 0.05? (Change parameter h and the number of iterations.)

Q5 Describe how the iterated image changed with this new step size.

You might already know the antiderivative of $\sin(x)$, so you could calculate $F(2)$ directly (always remembering the initial conditions), but in many cases it is not possible to find an expression for the antiderivative of a function, so your only option is Euler's method. Here's an example.

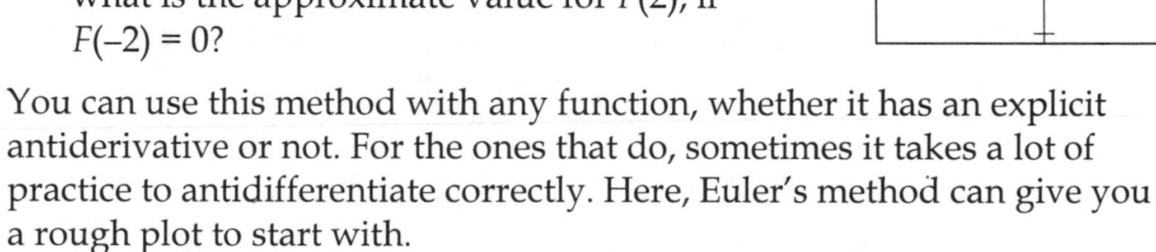

7. Double-click on the expression for $f(x)$ and change it to `f(x)=sin(sqrt(abs(x)))`.

You will need to change the value of the parameter y_R here.

Q6 Now, you can't find an expression for $F(x)$, but, using Euler's method with a step size of 0.05, what is the approximate value for $F(2)$, if $F(-2) = 0$?

You can use this method with any function, whether it has an explicit antiderivative or not. For the ones that do, sometimes it takes a lot of practice to antidifferentiate correctly. Here, Euler's method can give you a rough plot to start with.

8. Go to page 2 of the document. Here we have the same situation as on page 1, except that here point R is independent or free to move.

9. Select the measurements $x_R + h$ and $y_R + f(x_R) \cdot h$ in that order and choose **Plot as (x, y)** from the Graph menu. Label this point Q. With this new point, you can repeat Euler's method.

Make sure point Q is not selected as well by clicking once in an empty spot with your **Arrow** tool.

10. Select point R, and choose **Iterate** from the Transform menu. Click on point Q to match the pre-image (point R) with the first image (point Q). Uncheck **Tabulate Iterated Values** in the Structure pop-up menu. Click the *Iterate* button, and three iterations will be constructed.

11. With the new iterated image selected, choose **Edit | Properties.** Change the number of iterations to 200 on the Iteration panel.

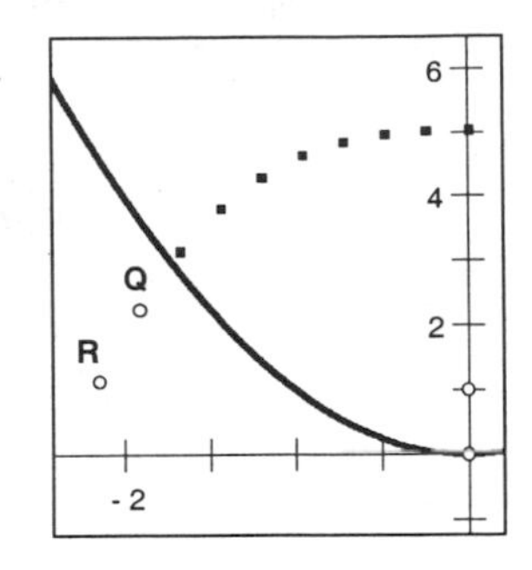

12. Use the slider to adjust the value of h to approximately 0.02.

You can enter your guess using the Graph menu and then see how well it matches the iterated image.

Q7 Do you recognize this plot? Can you guess its equation?

Exploration 1

On pages 3–7 there are various functions and their plots. For each one, an iterated image using Euler's method with step size h has already been constructed.

1. For each function $f(x)$, use the iterated image to predict the expression for the antiderivative. Write your prediction on a separate sheet of paper.
2. Test your prediction by choosing **Plot New Function** from the Graph menu and entering your expression. If you need some constants, press the *Show Sliders* button and use the sliders provided to build the expression.
3. If your prediction is not correct, compare the plot of your guess and the iterated image for hints as to how you should edit your function.
4. As you saw in step 7, there are functions for which it is not possible to find an antiderivative expression using elementary functions. Examine the function f on page 8. Iterate point R as you did before, and examine the image created.

Q1 Sketch your iterated image of the antiderivative that you constructed in step 4 on your paper. What is the approximate value for $F(5)$ using Euler's method with a step size of 0.05, if $F(-10) = -1.75$?

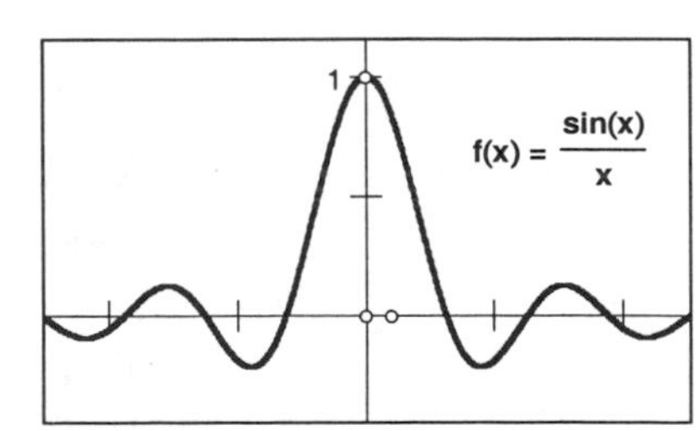

Exploration 2

You have seen that when a function is translated vertically, its derivative is unchanged. Is its antiderivative unchanged as well?

1. Go to page 9 of the document. On this page, $f(x) = \sin(x) + c$.

Q1 Adjust the slider for c. What happens to the antiderivative approximation, or the iterated image, when you translate the plot of $f(x)$ vertically upward? Vertically downward? Why does this happen?

Q2 In general, how will the antiderivative of $f(x)$ differ from the antiderivative of $f(x) + c$?

The sine function, its plot, and its antiderivative approximation are also useful for looking at other transformations.

2. Go to page 10 of the document. Here you'll find the function $f(x) = \sin(x)$, a transformation of $f(x)$ labeled $g(x)$, a couple of sliders, and the antiderivative approximations for each function constructed as before.

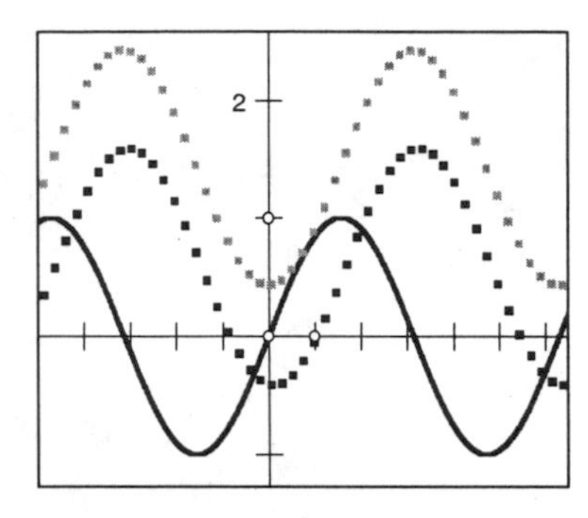

3. Adjust the slider for c and investigate how changing the value of c affects the approximated antiderivative plot of the function $g(x)$, or $f(x + c)$.

Q3 Are the antiderivatives of $f(x)$ and $f(x + c)$ horizontal translations of each other for all initial conditions? If not, can you adjust the initial conditions so that this is the case? Or is there some other relationship between the antiderivatives? If so, what is it?

4. Edit the function g so that $g(x) = c \cdot f(x)$. Investigate the relationships as you did in step 3.

Q4 Are the antiderivatives of $f(x)$ and $c \cdot f(x)$ vertical stretches of each other for all initial conditions? If not, can you adjust the initial conditions so that this is the case? Or is there some other relationship between the antiderivatives? If so, what is it?

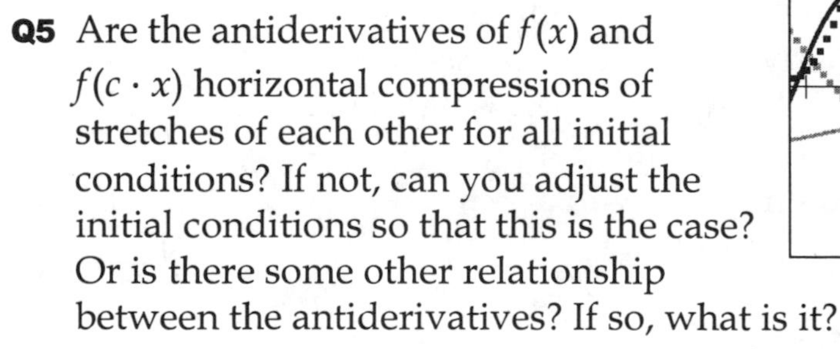

5. Edit the function g so that $g(x) = f(c \cdot x)$. Investigate the relationships as you did in step 3.

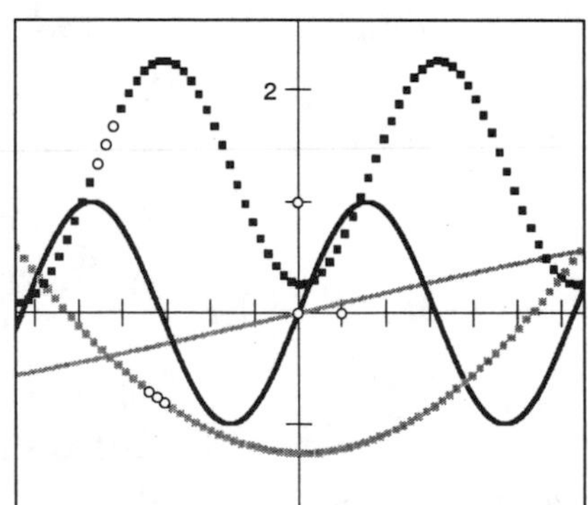

Q5 Are the antiderivatives of $f(x)$ and $f(c \cdot x)$ horizontal compressions of stretches of each other for all initial conditions? If not, can you adjust the initial conditions so that this is the case? Or is there some other relationship between the antiderivatives? If so, what is it?

A Field of Slopes

Name(s): ___________________________

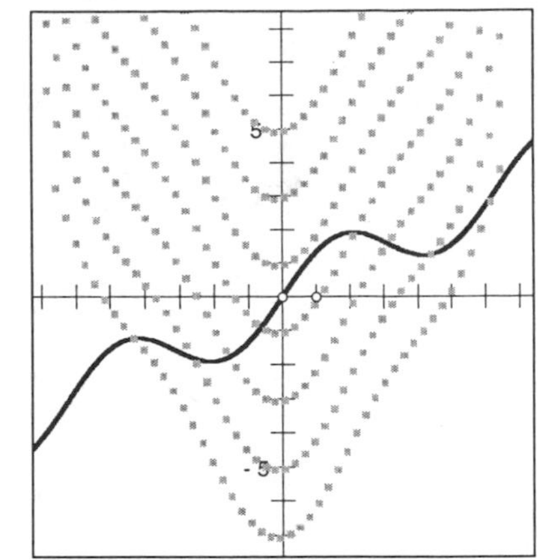

Finding the antiderivative $F(x)$ from the equation of $f(x)$ where $F'(x) = f(x)$ is called solving a *differential equation.* As you have seen, a differential equation has many solutions, because if $F(x)$ is an antiderivative of $f(x)$, then so is $F(x) + C$ for any constant C. So every differential equation actually describes a family of functions that are translations of each other.

In this activity, you will investigate *slope fields,* which are a method for visualizing the family of functions described by a differential equation, and apply a method for approximating a particular solution from the many possible solutions.

Sketch and Investigate

1. **Open** the document **Slopefield.gsp** in the **Exploring Antiderivatives** folder. On page 1 of this document you will find a function $f(x)$, its plot, an independent point R, and a slider.

Select point *R*, then choose **Measure | Abscissa (x).** (Choose **Ordinate (y)** for the *y*-coordinate.)

2. Because point R is independent, we can assume that it is on the antiderivative F. Measure the x- and y-coordinates of point R.

In the previous activities you learned that if $F(x)$ is a solution to the differential equation $F'(x) = f(x)$ and $F(x)$ contains point R, then at point R, the tangent line to F must have the equation $y_R + f(x_R)(x - x_R)$ and near point R, the antiderivative $F(x)$ must *look like* this tangent line.

3. Calculate $f(x_R)$ by choosing **Measure | Calculate.** Label this calculation *SlopeAtR* by double-clicking on it with your **Text** tool.

Click on the measurements to enter them into the Function editor.

4. To plot the tangent line, choose **Plot New Function** from the Graph menu and enter the expression $y_R + SlopeAtR \cdot (x - x_R)$.

5. Relabel this function T by double-clicking on the equation with your **Text** tool.

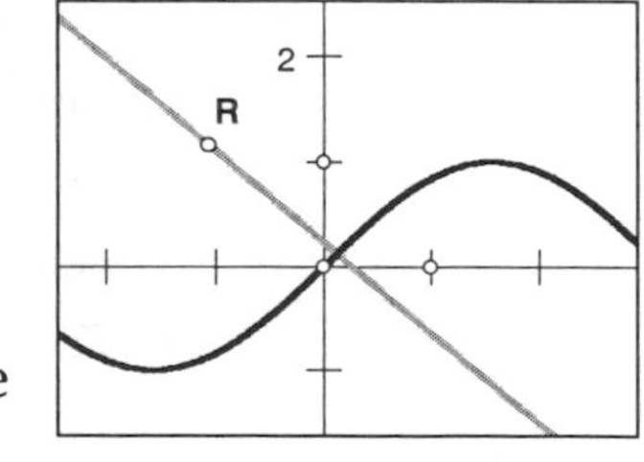

The tangent line only approximates the plot of $F(x)$ for a small interval around point R, so next you'll create a short segment of a fixed length around point R, with a slope equal to the slope of the tangent line by using the slider that adjusts the measurement s.

Click on the measurements to enter them into the calculator.

6. Choose **Measure | Calculate** and calculate $x_R + s$. Then calculate $T(x_R + s)$.

7. Select $x_R + s$ and $T(x_R + s)$ in that order, then choose **Plot As (x, y)** from the Graph menu. Label this point Q.

8. Plot the point $(x_R + s, y_R)$.

Or, using the **Circle** tool, you can click on point R, and then on the new point.

9. Construct a circle with center R and radius s by selecting point R and the new point from step 8 and choosing **Circle by Center+Point** from the Construct menu.

10. Select the tangent line and choose **Display | Hide Function Plot.** Then select points R and Q and choose **Construct | Line.**

You can also make a segment by selecting both points with the **Segment** tool.

11. Construct both intersections of the circle and the line RQ by clicking on the intersections with your **Arrow** tool. Then construct a segment between these two intersections by clicking on both points and choosing **Construct | Segment.**

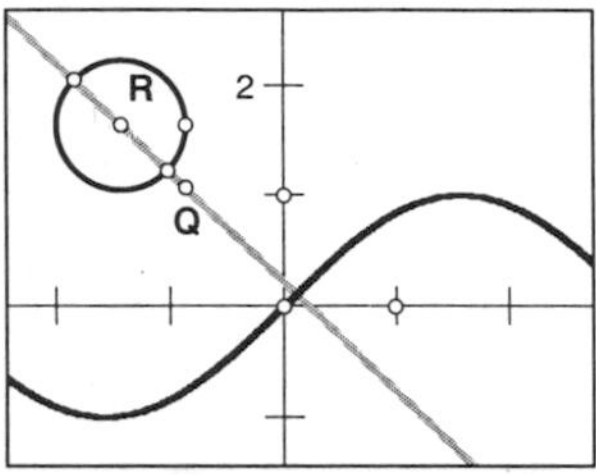

12. Hide the point $(x_R + s, y_R)$, the circle, and both intersection points by selecting them all and choosing **Display | Hide Objects.**

13. Adjust the slider for s to approximately 0.3.

The segment you've created centered at point R has slope $F'(x_R)$ or $f(x_R)$ and a fixed length. Now for the fun part! The segment at R shows you the slope at a single point. Dragging R shows the slope at *every* point of the drag. A *slope field* diagram describes the slope at many points simultaneously, giving a "big picture" of the shape of all solutions to the differential equation.

14. Select the segment and choose **Display | Trace Segment.** Make the segment thicker by choosing **Display | Line Width | Thick.** Using tracing here shows the segment's slope at various positions of R simultaneously.

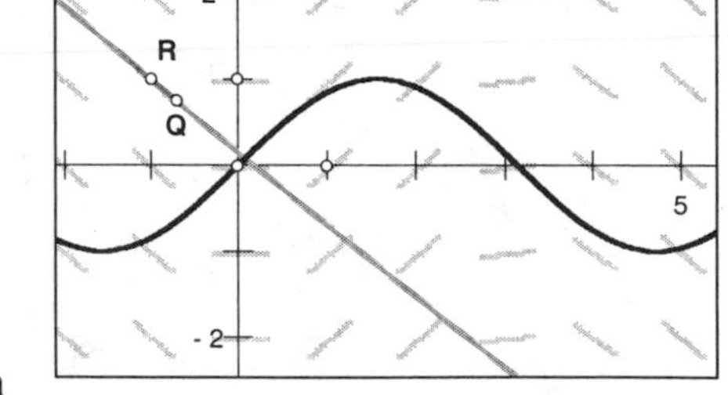

15. Choose **Snap Points** from the Graph menu. Snap points causes dragged points to stick to the grid points of your coordinate system.

16. Drag point R. As you drag, the traced segment shows the local slope for each point on the grid. Try to drag R so that you fill in the entire grid.

Q1 In your slope field, where is the segment's trace horizontal or closest to horizontal? Where is the trace closest to vertical?

Q2 Explain your answer to the previous question algebraically in terms of your differential equation. Why is the line horizontal or near-vertical in the cases you found in Q1?

A Field of Slopes (continued)

Q3 Suppose you dragged point R around the plane and traced the slope field in the picture on the right. What can you say about the differential equation?

Now, in order to find a particular solution to a differential equation, you must have at least one point that lies on $F(x)$—called an *initial point*. Suppose, for example, that $F(x)$ passes through the point $(-9, 7)$. You can now trace an approximate particular solution by "following" the tangent line.

17. Select point R, and then point Q, and choose **Edit | Action Buttons | Movement.** Click OK. This button will move point R toward point Q so that point R will *follow* the tangent line.

18. Move point R to the point $(-9, 7)$. Then press the *Move R→Q* button.

Q4 Given that the antiderivative of $\sin(x)$ is $-\cos(x) + C$, find the actual equation of the function that you've approximated by this trace. Plot your answer to see how it compares with the approximation.

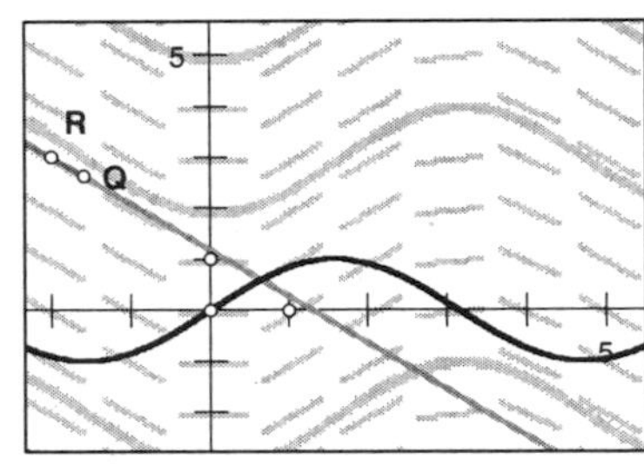

If point R moved off your screen, choose **Edit | Undo Animate Point** to bring it back to its starting point.

Q5 Move point R to another location and press the *Move R→Q* button. Find the equation for this new trace as you did in Q4. Is this trace a vertical translation of your answer in Q4?

Looking at the traces of the approximate particular solutions, you can see that the segment traces that make up the slope field suggest the slope or direction of all the possible solutions to a given differential equation.

Q6 Go to page 2 of the document. Here $f(x) = x$. Predict what the slope field will look like for this differential equation, and then create the field as you did in step 16 above.

If you don't have access to a printer, copy the diagram on page 109.

Q7 Once you've diagrammed the slope field for your differential equation, choose **File | Print** to print your sketch. Describe the shape and location of the patterns that occur in the slope field.

Q8 On your printout, use different colored pencils to trace out three particular solutions you see in your slope field.

In the questions above, you could probably predict what the slope field and the particular solution would look like from what you know about derivatives and antiderivatives. But slope fields can also be used for differential equations whose solutions aren't so obvious.

The alternative symbol for derivative is used here because the differential equation is a function of both x and y.

For instance, what function satisfies the differential equation $\frac{dy}{dx} = y$?

What does its slope field look like? Here the slope of the solution at a point depends on y instead of x. In other words, at any point (x, y), the solution has a tangent line with slope equal to the y-coordinate of that point. So, for example, at the point (2, 3), the solution $F(x)$ has a tangent line with slope 3. Is there a function with this property? You can use slope fields to find out.

The grid spacing can be adjusted by pressing the *Show Grid Controls* button and then editing the parameter *scale*.

19. Go to page 3 of the document. On this page you will find a point R and a segment centered at R, as on page 1. Here, though, point R is constrained to move along the locus of grid points so that you can create a slope field with more detail.

Q9 Use a piece of graph paper to sketch a slope field for this differential equation. At each point R, (x_R, y_R), draw a small segment whose slope is equal to the value of y_R.

20. To check your answer for Q9, select *SlopeAtR* and choose **Edit | Edit Calculation.** Then enter in the expression y_R for the new slope.

21. To create the slope field, move point R to every grid point.

If you don't have access to a printer, copy the diagram on page 109.

Q10 Print your sketch by choosing **File | Print.** Describe the shape and location of the patterns that occur in the slope field.

Q11 On your printout, use different colored pencils to trace out three different particular solutions you see in your slope field.

22. To check your answer to Q11, press the *Show Initial Pt* button and move point *Initial* to any one of your initial conditions in Q11.

23. Select point R and choose **Edit | Split Point from Locus** and press the *R→Initial Pt* button. Now press the *Move R→Q* button.

Q12 What function does your particular solution look like? Using your knowledge of derivatives, can you prove that it is this function?

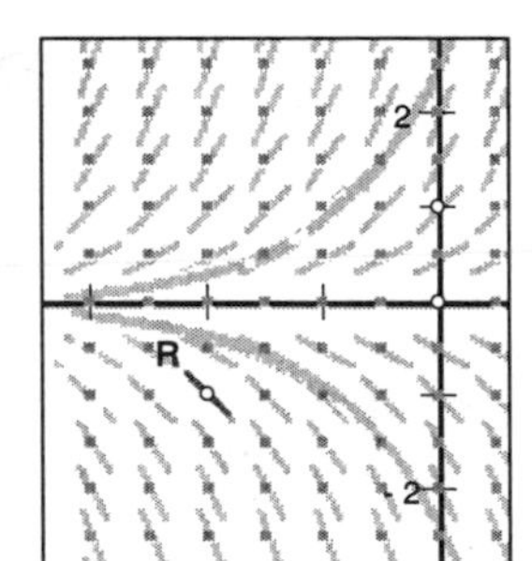

Explore More

For each differential equation answer these questions.

A. $\dfrac{dy}{dx} = 0.5 \cdot x$ B. $\dfrac{dy}{dx} = y - 2$

Q1 Describe how the slope field is similar to and different from $\dfrac{dy}{dx} = y$.

Select point R and any point in the grid, then choose **Edit | Merge Point To Locus.**

Q2 Repeat steps 19–23 and answer Q9–12 for each equation.

Stepping Through the Field

Name(s): ____________________

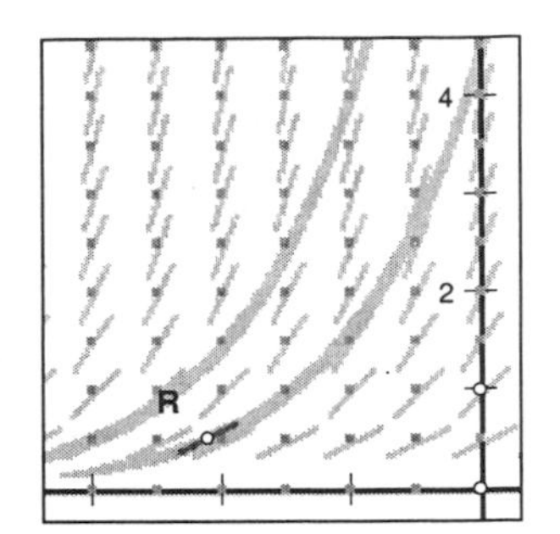

In the previous activity, you were introduced to slope fields as a way to visualize the family of solutions to a differential equation. You saw that slope fields could be used as a map—pick a starting point and follow the direction of the segments through the field.

Slope fields *point the way* along solution curves by showing the slope of the tangent to the solution curve. In this activity, you will see how Euler's method tells you *how* to move through a slope field numerically, step by step. You will also explore a wider variety of slope fields, use them as maps to draw possible solution curves, and apply Euler's method to approximate a particular solution numerically.

Sketch and Investigate

To start the process, let's examine the differential equation $\frac{dy}{dx} = \frac{x}{y}$.

What does its slope field look like? Here the slope of the solution at a point depends on both x and y, or in other words, at any point (x, y), the solution has a tangent line with slope x/y. So, for example, at the point (2, 3), the solution $F(x)$ has a tangent line with slope 2/3.

Q1 Use a piece of graph paper to sketch a slope field for this differential equation. At each point, $\{(x_R, y_R) \mid -3 \le x_R \le 3, -3 \le y_R \le 3\}$, draw a small segment with slope x_R/y_R.

The grid is really a locus of grid points, and point *R* is on the locus.

1. **Open** the document **Euler.gsp** in the **Exploring Antiderivatives** folder to check your answer. On page 1 of this document there is a grid of points. Point *R* is on the grid, and the segment centered at point *R* has been constructed so that its slope is equal to x_R/y_R.

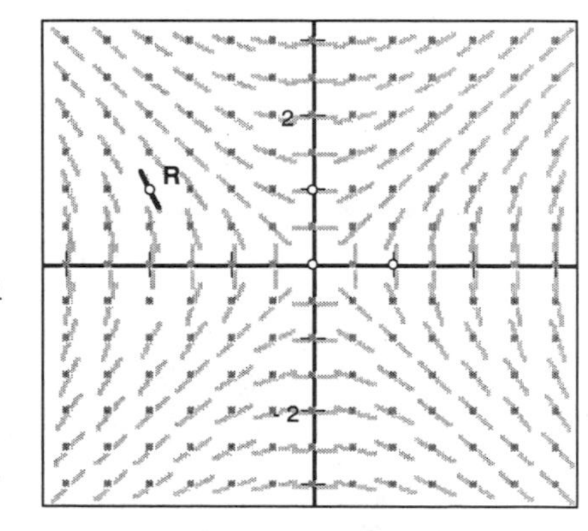

2. Check your slope field by dragging point *R* around the grid—make sure you have a trace at all the grid points.

If you don't have access to a printer, copy the diagram on page 109.

Q2 Choose **File | Print** to print your sketch. Describe the shape and location of the patterns that occur in the slope field.

Q3 On your printout, use different colored pencils to trace out three different particular solutions in your slope field.

3. To test one of your particular solutions, press the *Show Initial Pt* button and move point *Initial* to one of your initial points in Q3.

4. Select point R and choose **Edit | Split Point from Locus**. Press the *R→Initial Pt* button and then the *Show Point Q* button.

Adjust the slider labeled *Direction* to move point R in the opposite direction.

5. Now, press the *Move R→Q* button. Observe how point R follows the direction suggested by the segments and the slope field.

In "Plotting the Antiderivative," you saw that Euler's method is an iterative algorithm with these steps:

i. Start with an initial point or condition or (x_R, y_R).

ii. Calculate the slope at that point using the given equation—in this case, x_R/y_R.

iii. For the given value of h, calculate the change in y using this slope —specifically, $slope \cdot h$. Add this calculation to y_R to get a new point, $(x_R + h, y_R + slope \cdot h)$.

iv. Repeat this process with the new point.

Now, let's apply Euler's method using (–5, 4) as the initial point.

To remove clutter, you can hide point *Initial*'s label by clicking on it once with the **Text** tool.

6. Select point *Initial* and any grid point to select the locus of grid points. Choose **Edit | Merge Point To Locus.** Now move point *Initial* to the point (–5, 4). Press the *R→Initial Pt* button. The slope of the solution at this point is x_R/y_R, or –5/4.

Because h is set at 0.2, the y-value of the first approximation using Euler's method will be $y_R + slope \cdot h = 4 + (-1.25)(0.2) = 3.75$. So the first approximated point will be $(x_R + h, y_R + slope \cdot h)$, or (–4.8, 3.75).

7. Point Q has been constructed using Euler's method. Choose **Measure | Coordinates** to measure point Q's coordinates. Do the coordinates agree with the calculations above?

Check the status line to make sure that you have selected point R. If not, click on point R again.

8. Press the *Hide Segment* button. To see Euler's method repeated, select point R and then choose **Iterate** from the Transform menu. Click on point Q to match the image point. Set the number of iterations to 1 by pressing the minus (–) key twice. Uncheck **Tabulate Iterated Values** in the Structure pop-up menu. Click Iterate.

Q4 By hand, apply Euler's method to the point (–4.8, 3.75). What is the next approximated point?

The choice, **Terminal Point**, will only show up if you have selected only the last created point.

9. To check your answer to Q4, select the iterated image by clicking on the last created point, and choose **Transform | Terminal Point.**

10. With the terminal point selected, choose **Measure | Coordinates** to measure its coordinates. Do the coordinates agree with Q4?

Make sure the status line says "Selected: 1 iterated image". Otherwise, click on the point again.

11. Click on the iterated image and increase the number of iterations by pressing the plus (+) key.

Q5 What happens when you use Euler's method with an h of 0.2 after a total of 12 iterations (counting point Q as the first iteration)? Explain why this happens.

12. Select parameter h, choose **Edit | Edit Parameter**, and change its value to 0.1. Then, again select the iterated image, but this time, choose **Edit | Properties**. Change Number of Iterations to 19 on the Iteration panel. Click OK.

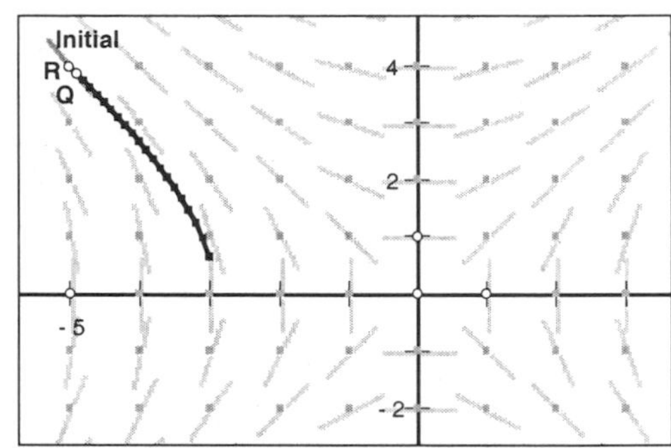

Q6 Your terminal point has coordinates (–3, 0.643). What do these coordinates mean in relation to the differential equation?

When you click on the location of the terminal point, the iterated image will be selected first.

Q7 How could you use your sketch to approximate $F(3)$ using Euler's method with initial point (1, 0.5) and step size $h = 0.1$? Try it.

You can use this document to experiment with all sorts of different differential equations by following the steps below.

13. Go to page 2 to experiment with differential equations that have a variable in the denominator. Examine the equation $\frac{dy}{dx} = \frac{x_R + y_R}{y_R^2 + 1}$ by editing the calculation *numerator* to equal $x_R + y_R$, and *denominator* to equal $y_R^2 + 1$. (The calculation *SlopeAtR* will be updated automatically.) Move point R around the grid to create a trace of the slope field.

14. Go to page 3 to experiment with differential equations that do not have a variable in the denominator. Examine the equation $dy/dx = (x_R - y_R)/3$ by editing the calculation *SlopeAtR* directly. Move point R around the grid to create a trace of the slope field.

Exploration 1

Pages 4 and 5 of the document have slope fields already constructed. These are hidden when you open the page so that you can trace the slope field yourself as you did above. Pick a couple of the differential equations below and use pages 4 and 5 to do the following steps. (Use page 4 to explore to explore differential equations with constant denominators and page 5 for differential equations with variable denominators.)

A. $dy/dx = ay^2$

B. $dy/dx = ay^2/x$

C. $dy/dx = a(y/x)$

D. $dy/dx = 0.5x + y$

E. $dy/dx = ax(1 - x)$

F. $dy/dx = ax/by$

G. $dy/dx = cx/(y^2 + x^2)$

H. $dy/dx = cy/(y^2 + x^2)$

Q1 For each, first predict what the slope field will look like by evaluating the differential equation at a few different points (x, y).

1. Then, edit the expression for the slope as described above in step 13 or 14 and make your slope field. Press the *Show Field* button, then use the sliders to vary the values of the parameters in the differential equations and observe any changes in the slope field.

If you don't have access to a printer, copy the diagrams on pages 109 and 110.

Q2 Print your sketch by choosing **File | Print.** Describe the shape and location of the patterns in the slope field.

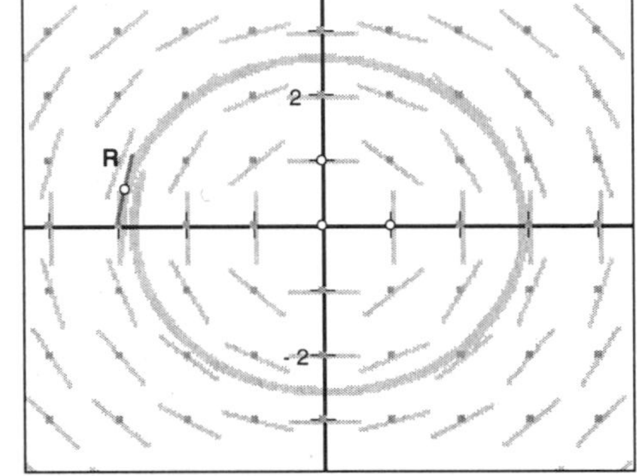

Q3 On your printout, use different colored pencils to trace out three different particular solutions in your slope field.

2. Create a particular solution by using the **Iterate** command as described above in step 8.

Q4 Describe how your particular solutions are similar and how they are different for different initial conditions.

Exploration 2

In Q5 of the Sketch and Investigate section, you saw that Euler's method can lead to poor approximations when vertical tangents are involved. Other methods for approximating solutions to differential equations are more precise.

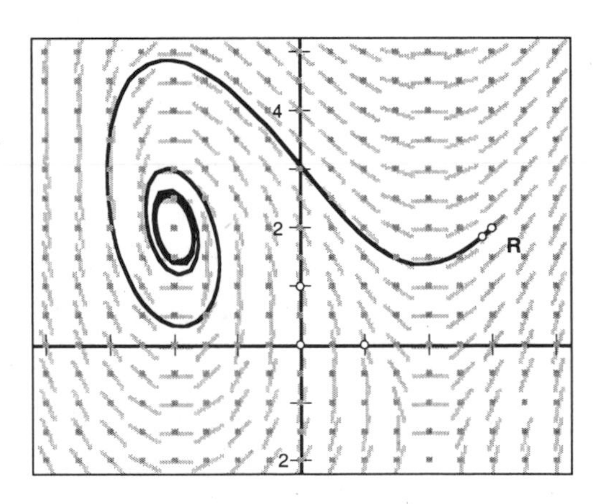

Q1 What is it in the differential equation on page 6 that causes the "whirlpool" effect in this solution curve?

A Field of Slopes:

Q7 $f(x) = x$

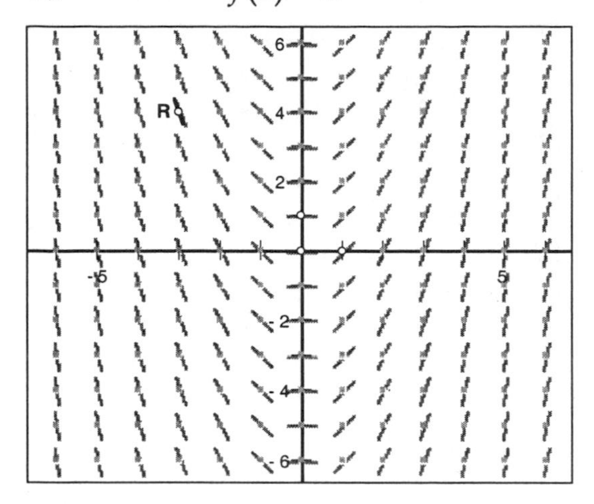

Q10 $dy/dx = y$

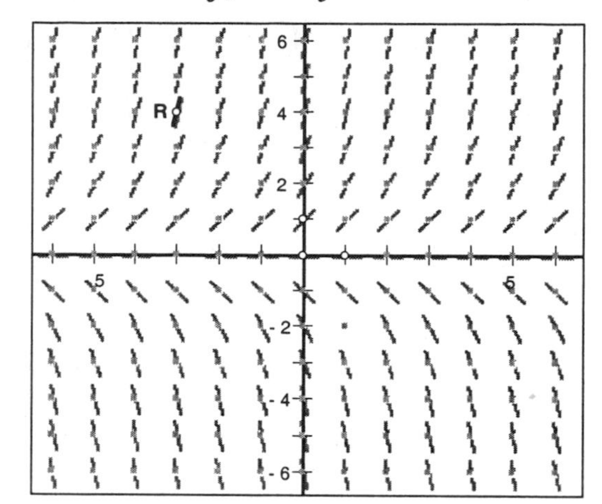

Stepping Through the Field:

Q2 $dy/dx = x/y$

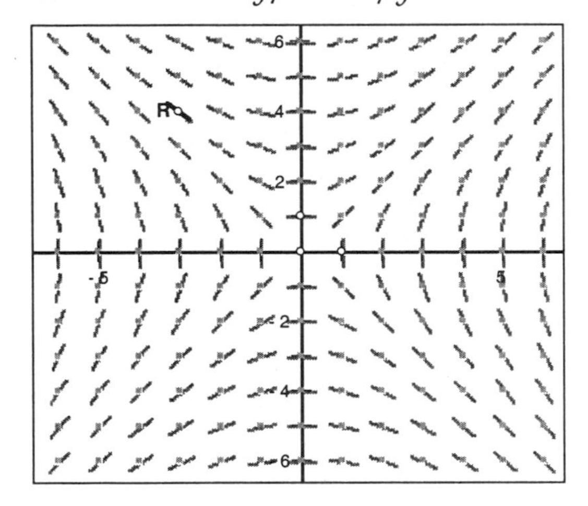

A. $dy/dx = ay^2$

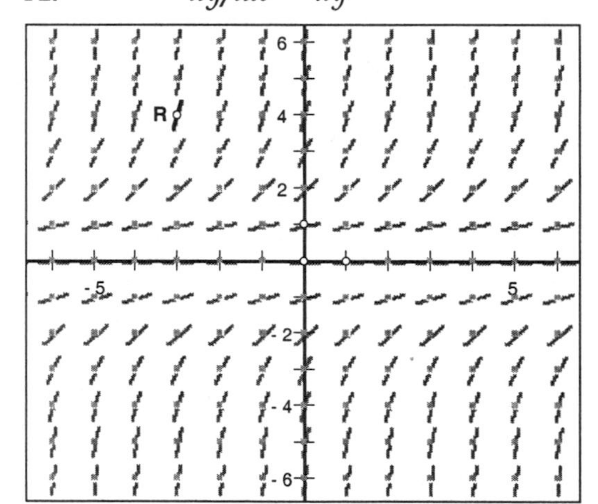

B. $dy/dx = ay^2/x$

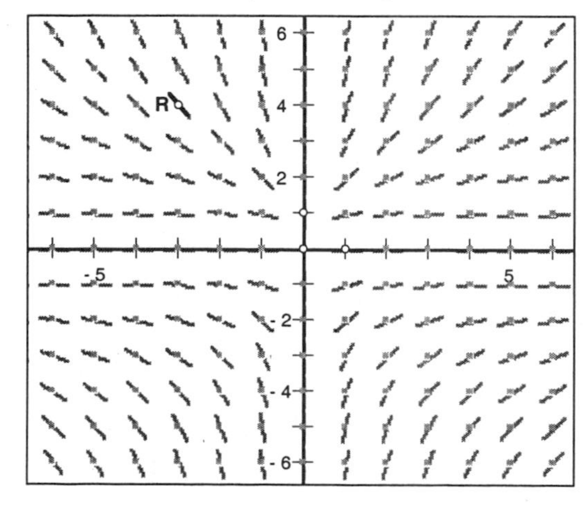

C. $dy/dx = a(y/x)$

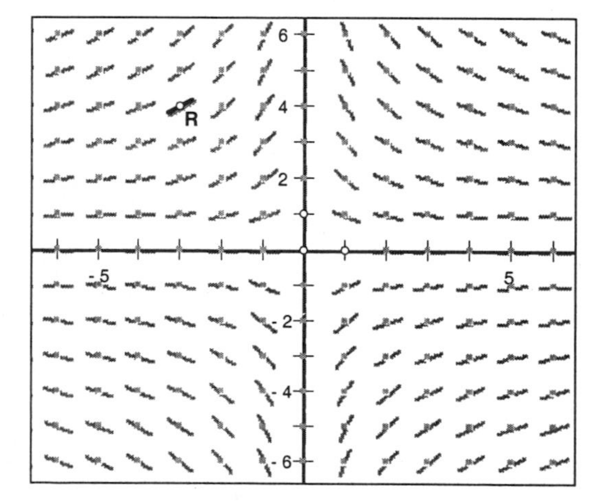

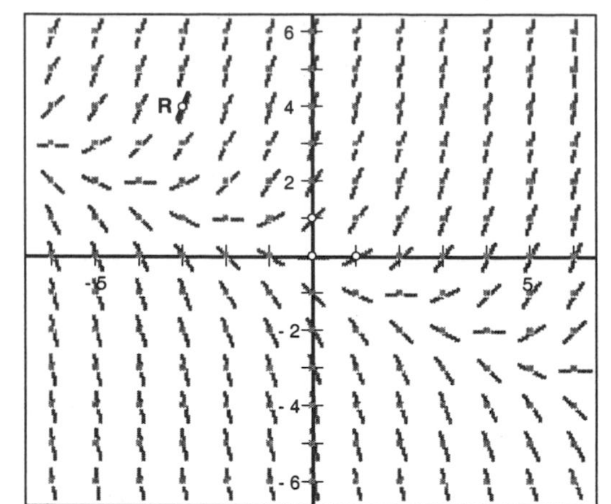

E. $dy/dx = ax(1-x)$

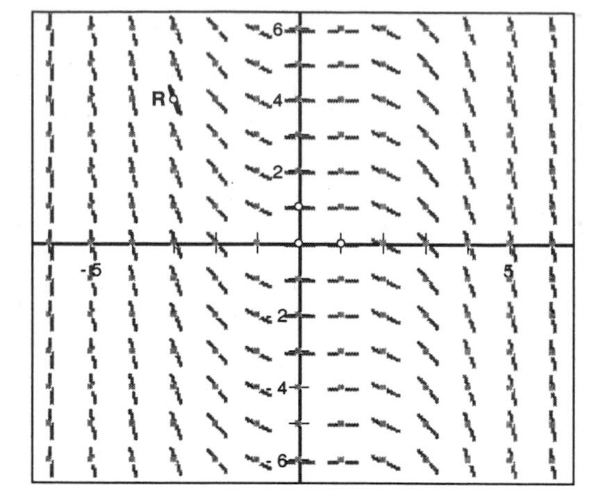

F. $dy/dx = ax/by$

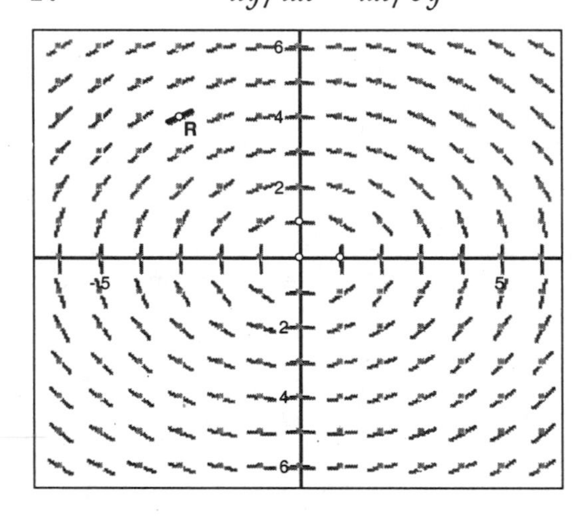

G. $dy/dx = cx/(y^2 + x^2)$

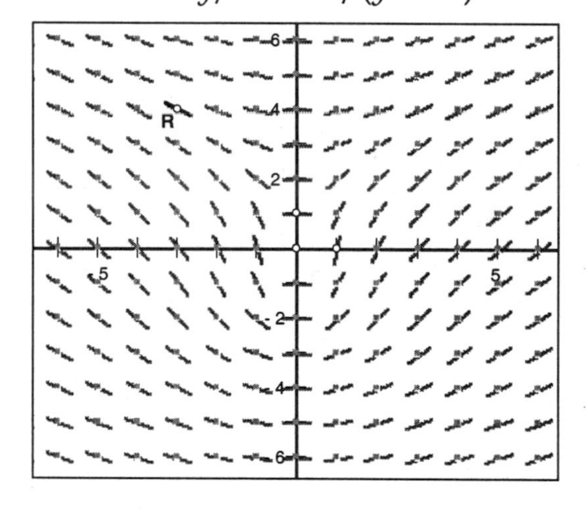

H. $dy/dx = cy/(y^2 + x^2)$

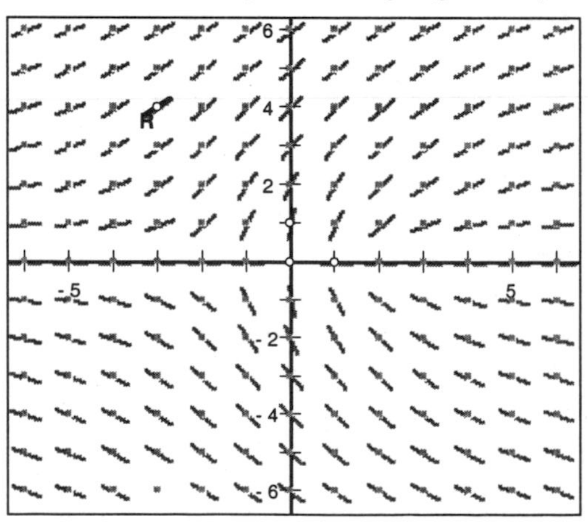

5

Exploring Integrals

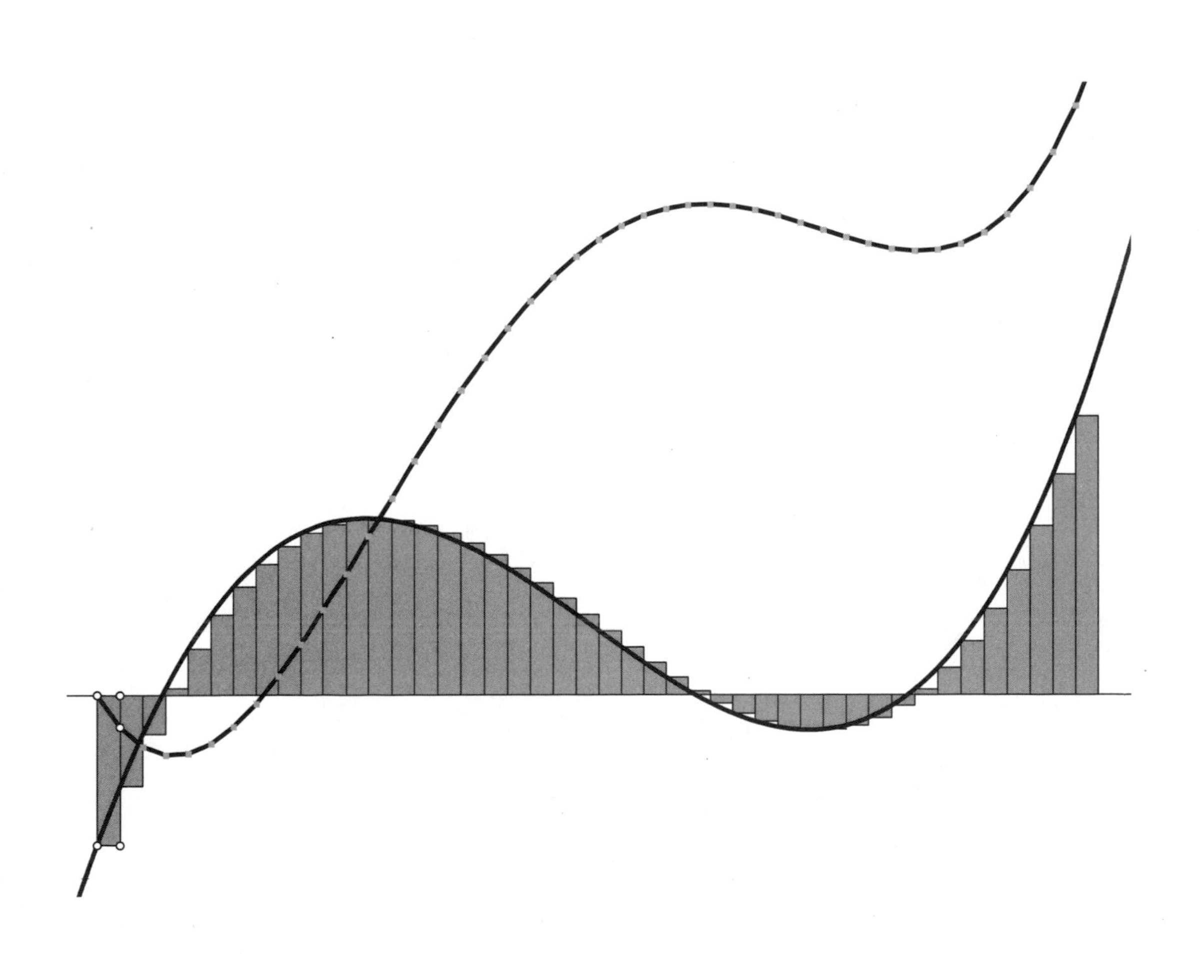

Building Area

Name(s): ______________________

Finding the area between the plot of a function and the x-axis has a surprising number of applications. You have seen that the area under a velocity curve is the object's change in position, or the net distance traveled over the given time interval.

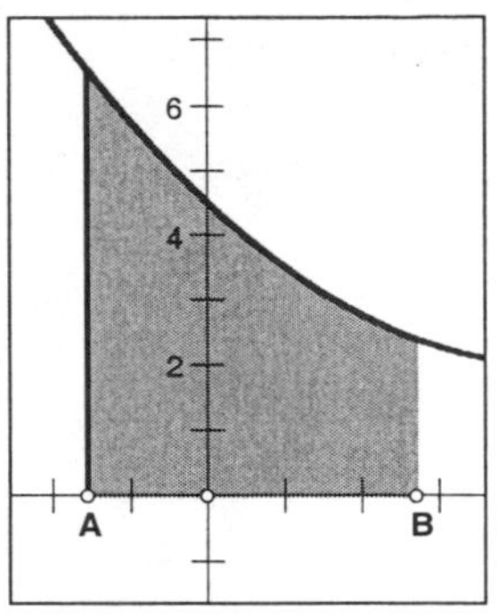

In this activity you will find the approximate area between a function plot and the x-axis using three different methods and tools. If you'd like to build these tools yourself, go to the Extension section of the Activity Notes after step 8.

Sketch and Investigate

1. **Open** the document **BuildArea.gsp** in the **Exploring Integrals** folder.

On page 1 of this document you will find the plot of a function $f(x)$, a slider that adjusts n, and an example of approximating the area between the plot of $f(x)$ and the x-axis from $x = 0$ to $x = 10$ by using rectangles.

Here n represents the number of rectangles.

2. Experiment with the slider that adjusts n. Make sure to try moving point N left and right of the segment that marks $n = 0$.

Q1 The rectangles are built differently, depending on whether point N is left or right of $n = 0$. Describe this difference.

Q2 Describe what happens to the rectangles, the area approximation, and the error as n increases.

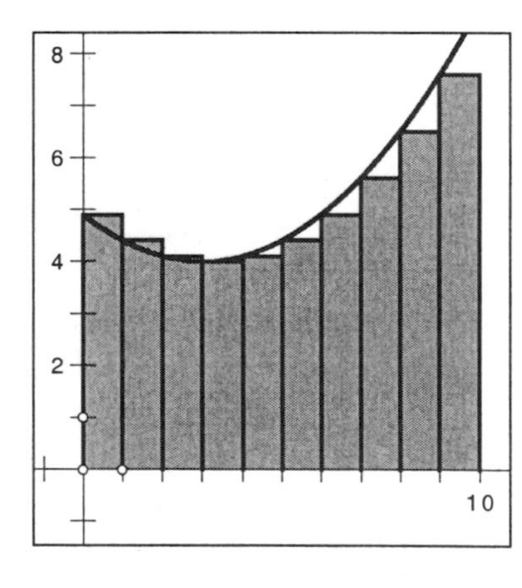

Using rectangles may underestimate or overestimate the area, but the smaller the widths, the better the total sum of the rectangles' area will approximate the actual area. In the steps below, you will build the rectangles yourself and find the sum of their areas on your own.

If a, b, or c is not as given in step 3, press the *Set Quadratic* button.

3. Go to page 2 of the document. Here $f(x) = a(x - b)^2 + c$, where $a = 0.5$, $b = 3$, and $c = 2$.

4. To make an interval on the x-axis, choose the **Point** tool and construct two points on the x-axis at approximately $x = -0.5$ and $x = 5.5$. Label these points A and B.

5. Select both points and then measure their x-coordinates by choosing **Measure | Abscissa (x).**

Click on measurements to enter them into the calculator.

6. Choose **Measure | Calculate** and calculate the value $x_B - x_A$. This will be the width of the interval over which you will construct rectangles.

7. To divide the interval from point A to point B into n equal subdivisions, enter $(x_B - x_A)/n$ in the calculator. With the **Text** tool, label the result h. This will be the width of each of the n rectangles.

8. Choose **Left Rectangles** from **Custom** tools. Then click on point A.

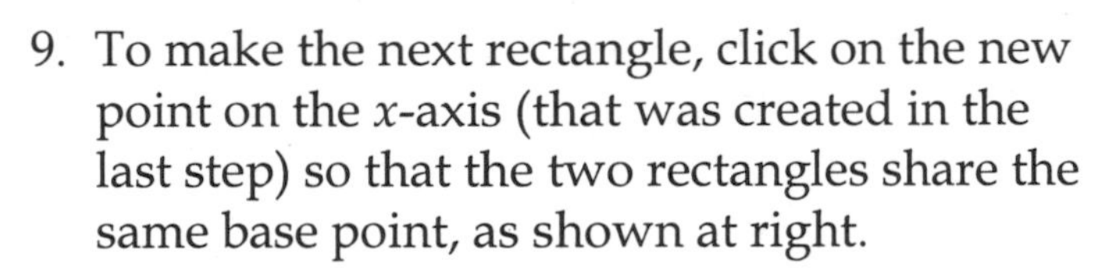
The tool will create a rectangle with the function's value on the left side of the subdivision as its height. The sum of the areas of a series of such rectangles is called a *left sum*.

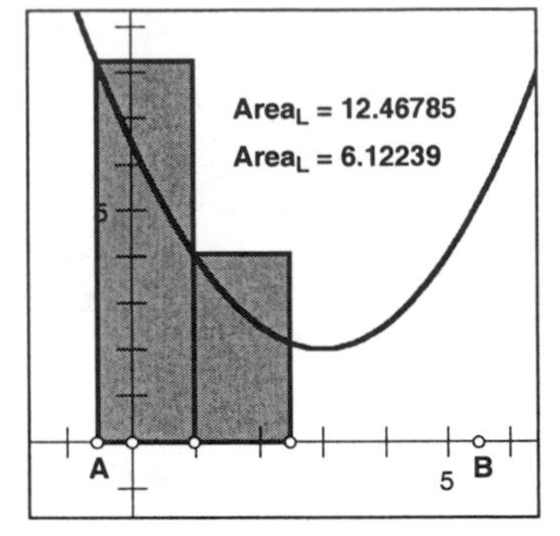

9. To make the next rectangle, click on the new point on the x-axis (that was created in the last step) so that the two rectangles share the same base point, as shown at right.

10. Continue to make rectangles until you have covered the interval from point A to point B. When you are done, you should have n rectangles.

11. Find the sum of the n area calculations by choosing **Measure | Calculate** and then entering each $Area_L$ measurement. Label this measurement *LeftSum* and save this calculation.

12. The area calculated by the tool is the product of the width of the rectangle (h) and the height (the value of the function at the left endpoint of the subdivision). Adjust the slider for c to translate the function vertically. Watch the area calculations as you raise and lower the function, especially when the function takes on negative values.

Q3 Why are some of your $Area_L$ measurements now negative? What can you do to make the measurement *LeftSum* negative?

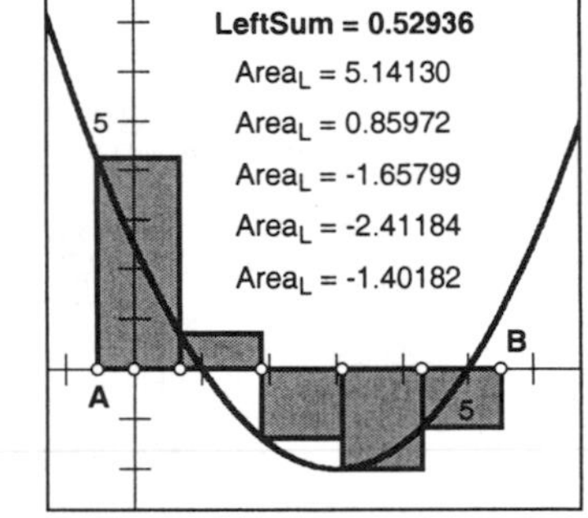

Q4 Given the equation of the function and a calculator, how would you produce the area measurements yourself?

Allowing negative areas (something you may never have considered before!) has both theoretical and practical uses, as you will see later on.

Next, you will create a *right sum*.

If needed, press the *Reset* button to reset the domain to [−0.5, 5.5].

13. Go to page 3 of the document. On this page you will find the same function that was on page 2, points A and B, and the calculation h. Points A and B are approximately in the same positions on the x-axis as they were originally on page 2.

14. Choose **Right Rectangles** from **Custom** tools. Then click on point A.

Q5 How is a right sum rectangle different from a left sum rectangle?

Here, the values for the sliders should still be $a = 0.5$, $b = 3$, and $c = 2$. If not, press the *Reset* button.

15. Continue to make rectangles as you did above in step 10. Make sure your value for n is the same as it was on page 2.

16. Find the sum of the n calculations of $Area_R$. Label it *RightSum* and save this calculation as well.

You wrote down your left sum in step 11.

Q6 How does the right sum compare with the left sum? Is one an underestimate and the other an overestimate?

17. Press the *Show Area* button to see both the exact region between the plot of $f(x)$ and the x-axis and its exact area.

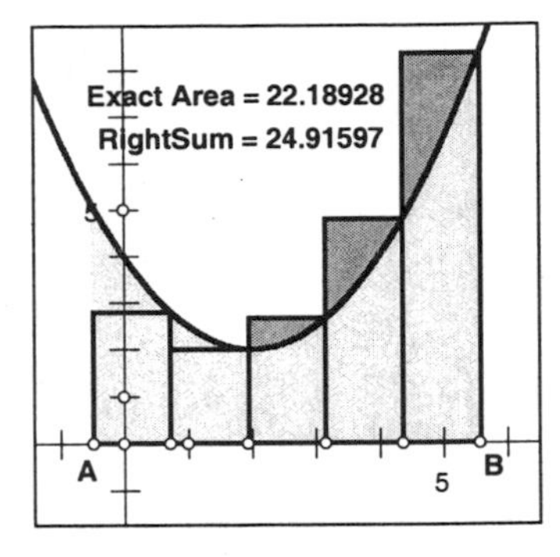

Q7 How does the right sum compare with the exact area? How does the left sum compare?

Remember that $\%Error = (RS - exact)/exact$, where *RS* means *RightSum* and *exact* is the actual area.

Q8 Use the calculator to find the percentage error for the right sum and write your result in the margin.

You can modify the function by adjusting the sliders for a, b, or c, or changing the interval $[x_A, x_B]$ by dragging point A or point B. Observe how the right sum and the exact area vary for different functions or on different intervals. Can you find a function where the right sum is always an overestimate? An underestimate?

You can also increase the value of n, but you will have to add rectangles and recalculate the sum. As you saw in Q2, increasing the number of rectangles gives a more accurate estimate of the area.

To edit the *RightSum*, double-click on the calculation and add in the new $Area_R$ measurements.

18. Increase the number of subdivisions by adjusting the slider for n. Fill in the rest of the interval by adding the rectangles you need using the **Right Rectangles** tool. Find the new sum.

Q9 How much did your percentage error drop?

Explore More

As you saw above, increasing the number of rectangles gives a more accurate estimate of the area. Another way to improve accuracy for your area estimate is by using trapezoids or midpoint sums. In the steps below, you will first build trapezoids, then another type of rectangle.

Check that this is indeed the function and domain. If not, press the *Reset* button.

1. Go to page 4 of the document. Here, you will be estimating the area under the curve $f(x) = 0.5(x - 3)^2 - 2$ on the interval $[-0.5, 5.5]$ again.

Make sure n is the same value you used back in steps 11 and 16.

2. Choose **Trapezoids** from **Custom** tools. Use this tool in the same way you used the other tools above to construct n trapezoids on the given interval and then calculate the approximate sum of the n trapezoids you constructed.

Q1 Looking at one trapezoid, what is the formula for the area using h and the function's values?

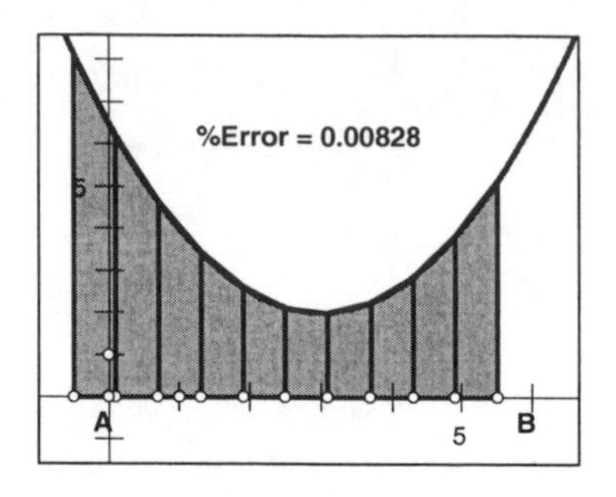

Q2 You can see graphically that the trapezoid sum gives a more accurate estimate of the area. Press the *Show Area* button and calculate the percentage error using trapezoids. Save this answer.

3. Press the Shift key, and then choose **Undo All** from the Edit menu.
4. Choose **Midpoint Rectangles** from **Custom** tools. Use this tool in the same way you used the other tools above to construct n rectangles on the given interval and then calculate the approximate sum of the n rectangles you constructed.

Q3 Looking at one rectangle, what is the formula for the area using h and the function's values?

Q4 Visually, how does it compare with the trapezoid or other rectangle tools? Press the *Show Area* button and calculate the percentage error using this tool. Save this answer.

Q5 Compare the various methods and the area approximations and percentage errors that you found back in Q8 and then in Q2 and Q4 above. If you had to make the tool, which would be the easiest to construct? Which is the most accurate? The least accurate?

The Trapezoid Tool

Name(s): ____________________

Using trapezoids and rectangles to approximate the area under a curve is a relatively easy way to go. The hard part is really not hard, just time-consuming. Subdividing your region into smaller regions, then constructing the trapezoids or rectangles for every single piece and measuring their areas—that can get tedious. In the last activity, you had a tool that easily made trapezoids and measured their area all at once. Building tools is what this activity is all about.

Sketch and Investigate

1. **Open** the document **TrapezoidTool.gsp** in the **Exploring Integrals** folder. You will be looking at the region on the interval [0, 2].
2. Using the **Point** tool, construct two points on the x-axis on the interval [0, 2]. Label the left point R and the right point Q.
3. Measure their x-coordinates by selecting point R and point Q, and then choosing **Abscissa (x)** from the Measure menu.

Now you need the top of your trapezoid to be anchored on the function at $f(x_R)$ and $f(x_Q)$.

4. Calculate the function values $f(x_R)$ and $f(x_Q)$ by choosing **Measure | Calculate,** then select the expression for f and the measurement x_R to enter them into the calculator. Do the same for point Q.
5. Select x_R and $f(x_R)$ in that order and choose **Graph | Plot as (x, y).** Plot the point $(x_Q, f(x_Q))$ as well.

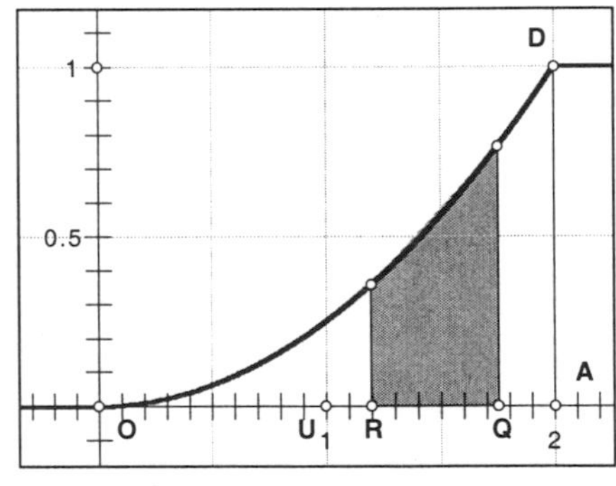

If you want to, you can also construct the sides by choosing the **Segment** tool and then selecting the two points of each side.

6. Select the four points of the trapezoid and choose **Construct | Quadrilateral Interior** to construct the interior of the trapezoid.

Now you have the actual trapezoid, but you need an area measurement as well. The base is the distance between point R and point Q, $x_Q - x_R$. The heights are the y-values, $f(x_R)$ and $f(x_Q)$.

7. Calculate the trapezoid's area by choosing **Measure | Calculate** and entering $0.5 \cdot (x_Q - x_R) \cdot (f(x_R) + f(x_Q))$. (Why does this formula work?)
8. Label this measurement *AreaT* and check Use Label In Custom Tools on the Label panel.

To make a tool, you need to choose the objects that are necessary to construct your final goal and you need to select the objects you want to end up with.

The Trapezoid Tool (continued)

If you constructed the line segments in step 6, select each of those as well.

9. Select the givens: point R, point Q, and the expression for $f(x)$. Then select the results: the region's interior, the points on the function, and measurement *AreaT*. Choose **Create New Tool** from **Custom** tools and name this tool **Trapezoid.** Check Show Script View and click OK.

After you're done, function f will be assumed.

10. In the tool's Script View, double-click given Function f and check Automatically Match Sketch Object. Click OK.

Choose **Edit | Undo Trapezoid** until you are back to your original trapezoid.

Now you have a trapezoid tool, so all you need to do to make a new trapezoid is choose your tool and make two points on the x-axis. Try it! When you're done experimenting, undo all trapezoids except for your original one.

11. Drag point R to the origin and point Q anywhere between 0 and 1.

12. Choose **Trapezoid** from **Custom** tools. Match point Q and click on the axis to construct a new point anywhere on the x-axis to the right of point Q except the unit point—point U.

13. Continuing with your **Trapezoid** tool, match the new point you constructed in step 12, and construct a new point anywhere on the x-axis to the right of that point. (Make each new trapezoid a different color.)

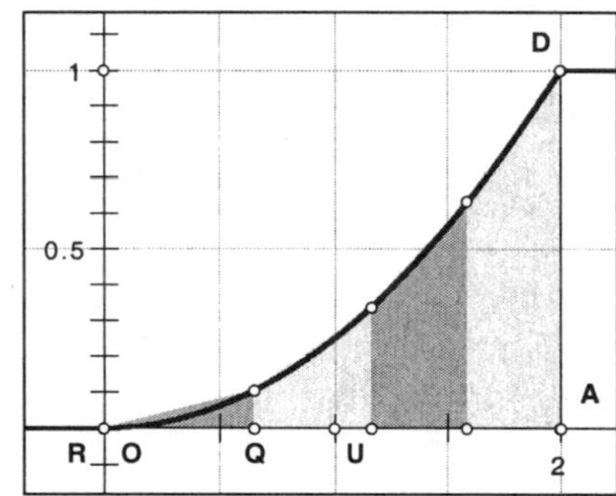

14. Do step 13 one more time, matching the previous point and then constructing a new point to the right of that, but this time, do *not* match point A. (You want the freedom to move your trapezoids, and point A, like point U, cannot move.)

15. You now have four trapezoids linked together. Move the points on the x-axis so that these four trapezoids completely cover the interval [0, 2].

16. Choose **Calculate** from the Measure menu and sum all the trapezoid measurements, *AreaT*. Label this measurement *Approximation*.

Q1 What is your approximation for the total area using four trapezoids?

17. Press the *Show Area Tools* button, then the *Reset P* button, and then the *Calculate Area* button to calculate the actual area under the curve and to shade in that region.

Q2 What is the actual area under the curve from $x = 0$ to $x = 2$?

18. Calculate your approximation percentage error:

$$\%Error = \frac{|Actual\ Area - Approximation|}{Actual\ Area}$$

Q3 Move the points on the x-axis to find the four trapezoids that give the smallest percentage error. What x-values did you find?

Exploration 1

Check the status line to make sure that you have not selected point *P*. If you have, click on the point again.

1. Erase traces and slide your four trapezoids to the left so they are all *inside* the region from $x = 0$ to $x = 1$, and then use your **Trapezoid** tool to construct four more trapezoids, linking them as before.

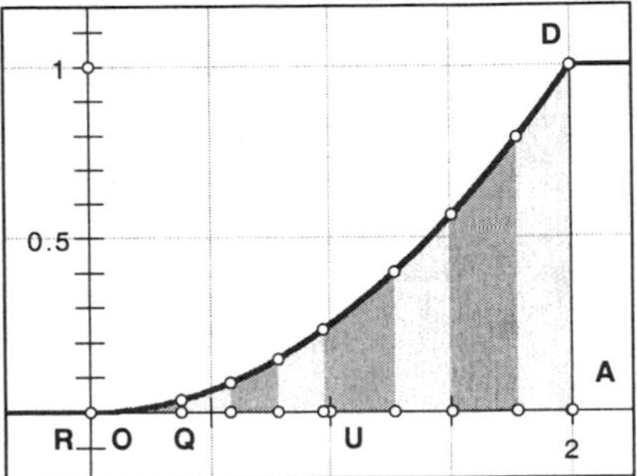

2. Once you have eight trapezoids, move them so that they cover the interval from $x = 0$ to $x = 2$.
3. With the **Arrow** tool, double-click on the measurement *Approximation,* and add the four new *AreaT* measurements. (So *%Error* will automatically update itself.)

Q1 What is your approximation for the area using eight trapezoids?

Q2 Move the points on the x-axis to find the eight trapezoids that give the smallest percentage error. What x-values did you find?

Exploration 2

Go to page 2 and experiment with your **Trapezoid** tool on this page. In particular, try each of the following constructions.

A. Construct at least one trapezoid in an interval where the function is below the x-axis during that whole interval.

B. Construct at least one trapezoid with one point in a region where f is below the x-axis ($f < 0$) and the other point in a region where f is above the x-axis ($f > 0$).

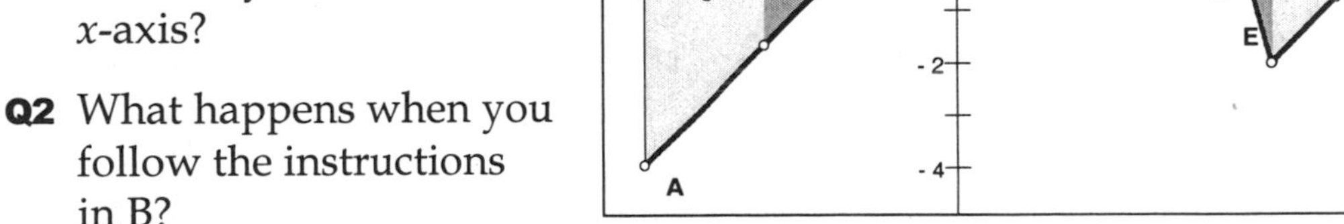

Q1 What happens when you use your **Trapezoid** tool on a part of the curve that is entirely below the x-axis?

Q2 What happens when you follow the instructions in B?

Q3 Cover the region with trapezoids to get an approximation for the total area under the curve, and then compare it with the "real" area using the **Area** tools. What values did you get?

Accumulating Area

Name(s): ____________________

How would you describe the shaded region shown here? You could say: The shaded region is the area between the x-axis and the curve $f(x)$ on the interval $0 \le x \le 4$. Or, if you didn't want to use all those words, you could say: The shaded region is

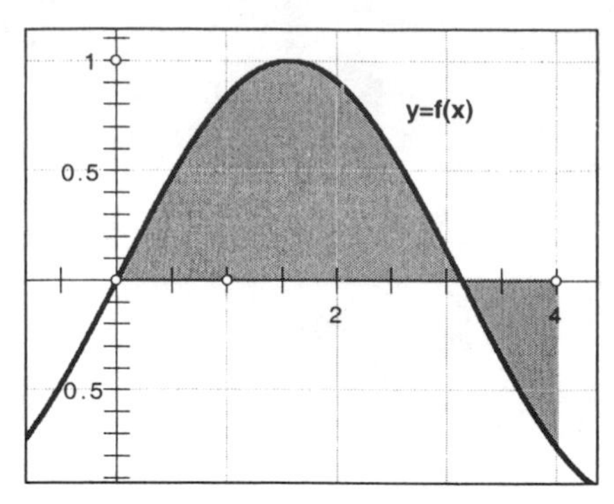

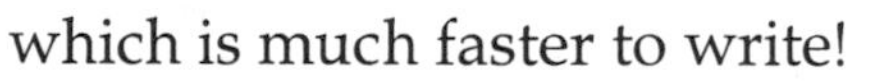

$$\int_0^4 f \quad \text{or} \quad \int_0^4 f(x)\,dx$$

which is much faster to write!

In general, the notation $\int_a^b f(x)\,dx$ represents the *signed area* between the curve f and the x-axis on the interval $a \le x \le b$. This means that the area below the x-axis is counted as negative. This activity will acquaint you with this notation, which is called the *integral,* and help you translate it into the signed area it represents.

Sketch and Investigate

1. **Open** the document **Area2.gsp** in the **Exploring Integrals** folder. You have a function f composed of some line segments and a semicircle connected by moveable points.

If you need to evaluate the integral $\int_0^4 f(x)\,dx$, the first step is to translate it into the language of areas. This integral stands for the area between f and the x-axis from $x = 0$ to $x = 4$, as shown. This area is easy to find—you have a quarter-circle on $0 \le x \le 2$ and a right triangle on $2 \le x \le 4$.

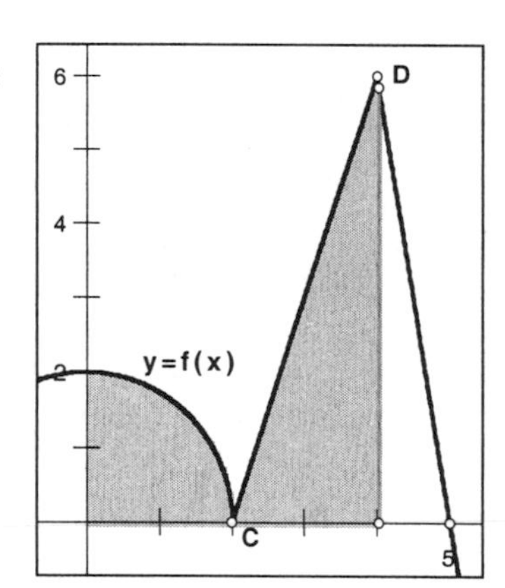

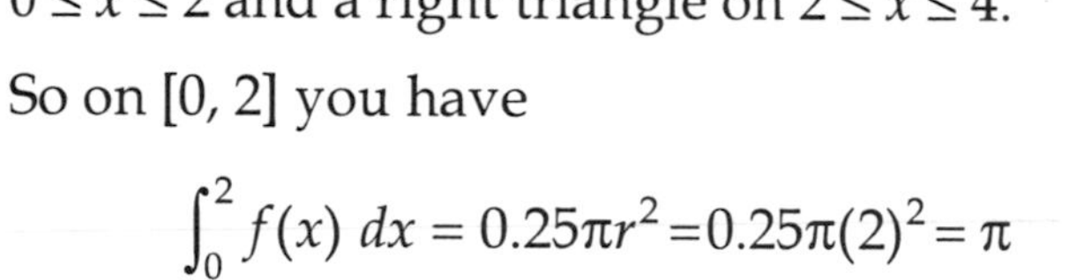

So on [0, 2] you have

$$\int_0^2 f(x)\,dx = 0.25\pi r^2 = 0.25\pi(2)^2 = \pi$$

and on [2, 4] you have

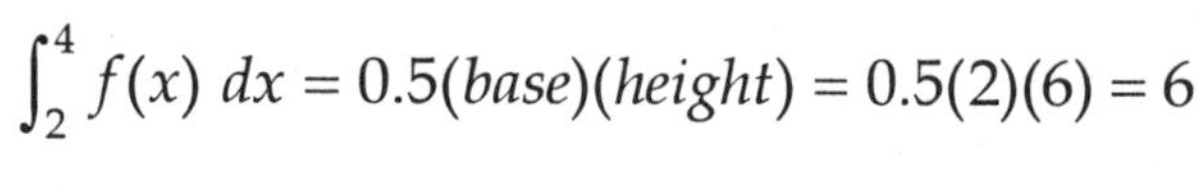

$$\int_2^4 f(x)\,dx = 0.5(base)(height) = 0.5(2)(6) = 6$$

so $\quad \int_0^4 f(x)\,dx = 6 + \pi$

2. To check this with the **Area** tools, press the *Show Area Tools* button.

There are three new points on the x-axis—points *start, finish,* and P. Points P and *start* should be at the origin. Point P will sweep out the area under the curve from point *start* to point *finish*. Point P has not moved yet, so the measurement *AreaP* is 0.

3. Press the *Calculate Area* button to calculate the area between f and the x-axis on the interval [*start, finish*] and to shade in that region.

Q1 Is the value of the measurement *AreaP* close to $6 + \pi$? Why isn't it exactly $6 + \pi$ or even 9.142?

Q2 Based on the above reasoning, evaluate $\int_{-2}^{4} f(x)\,dx$.

Before you move the point, check the status line to make sure you have selected the right point. If you haven't, click on the point again.

4. To check your answer, move point *start* to point B, press the *Reset P* button, and then press the *Calculate Area* button.

Q3 What do you think will happen to the area measurement if you switch the order of the integral, in other words, what is $\int_{4}^{-2} f(x)\,dx$?

5. To check your answer, move point *start* to $x = 4$ and point *finish* as close as you can get to $x = -2$, then press the *Reset P* button.

6. Choose **Erase Traces** from the Display menu and then press the *Calculate Area* button.

Q4 What is the area between f and the x-axis from $x = 4$ to $x = -2$?

Now, what happens if your function goes below the x-axis? For example, suppose you want to evaluate $\int_{4}^{6} f(x)\,dx$.

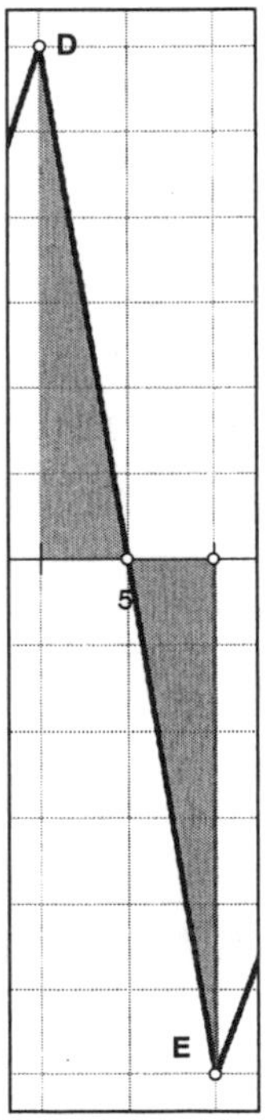

Q5 Translate the integral into a statement about areas.

Q6 What familiar geometric objects make up the area you described in Q5?

The grid is shown here for comparison. It doesn't appear in the sketch.

Q7 Using your familiar objects, evaluate $\int_{4}^{6} f(x)\,dx$. (*Hint:* You can do this one quickest by thinking.)

7. Make sure point *start* is at $x = 4$ and move point *finish* to $x = 6$. Press the *Reset P* button.

You can also erase traces by pressing the Esc key twice.

8. Choose **Erase Traces** from the Display menu, and then press the *Calculate Area* button to check your answer. Does the result agree with your calculation?

Q8 Evaluate $\int_{-6}^{-3} f(x)\,dx$ using the process in Q5–Q7 and check your answer using steps 7 and 8.

If you fix your starting point with $x_{start} = -6$, you can define a new function, $A(x_P) = \int_{-6}^{x_P} f(x)\,dx$, which accumulates the signed area between f and the x-axis as P moves along the x-axis.

Q9 Why is $A(-6) = \int_{-6}^{-6} f(x)\,dx = 0$?

Q10 What is $A(-3)$?

To get an idea of how this area function behaves as point P moves along the x-axis, you'll plot the point $(x_P, A(x_P))$ and let Sketchpad do the work.

9. Move point *start* to $x = -6$ exactly. Now move point *finish* to $x = 9$. The measurement *AreaP* is now the function $A(x_P) = \int_{-6}^{x_P} f(x)\,dx$.

To turn off tracing, choose **Trace Segment** from the Display menu.

10. Select the line segment that joins point P to the curve. Turn off tracing for the segment. Erase all traces.

If you can't see the new point, scroll or enlarge the window until you do. To enlarge the window, press the *Show Unit Points* button, resize the window, then press the *Set Function* button.

11. Select measurements x_P and *AreaP* in that order and choose **Plot as (x, y)** from the Graph menu.
12. Give this new point a bright new color from the Color submenu of the Display menu. Turn on tracing for this point and label it point I.
13. Press the *Reset P* button and then the *Calculate Area* button to move point P along the x-axis and create a trace of the area function.

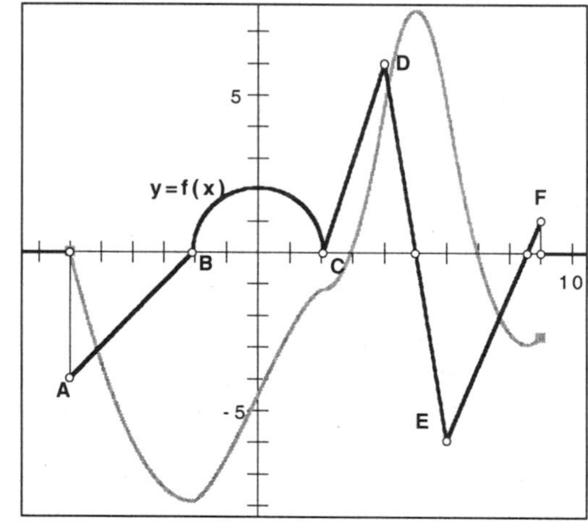

Q11 Why does the area trace decrease as soon as point P moves away from point *start*?

Q12 Why doesn't the trace become positive as soon as point P is to the right of point B?

Q13 What is the significance (in terms of area) of the trace's first root to the right of point *start*? The second root?

Q14 What is significant about the original function f's roots? Why is this true?

14. Turn off tracing for point I and erase all traces.

Check the status line to see that point P is selected. If not, click on the point again.

15. Select points P and I and choose **Locus** from the Construct menu.

The locus you constructed should look like the trace you had above. The advantage of a locus is that if you move anything in your sketch, the locus will update itself, whereas a trace will not.

Be sure to keep points A, B, C, D, E, and F lined up in that order from left to right. If point C moves to the right of point D, the line segment CD will no longer exist.

There are quite a few familiar relationships between the original function f and this new locus—including the ones suggested in Q11–Q14. See if you can find some of them by trying the experiments below.

A. Move point B (which also controls point C) to make the radius of the semicircle larger, then smaller.

B. Press the *Set Function* button to move point B back to (–2, 0). Now move point A around in the plane. (Make sure to stay to the left of point B.) Try dragging point A to various places below the x-axis, and then move point A to various places above the x-axis.

C. Press the *Set Function* button to move point A back to (–6, –4). Now move point D around in the plane. (Make sure to stay between point C and point E.) Drag point D to various places above the x-axis, and then drag point D to various places below the x-axis.

D. Follow step C with points E and F.

Q15 List the various patterns that you found between the two functions or in the area function alone. How many patterns were you able to find? Any conjectures about the relationship between the two functions?

Explore More

Will the area function's shape change if you move point *start* to a value other than $x = -6$?

1. Select point *start* and move it along the x-axis.

Q1 Does the area function's shape change when your starting point is shifted along the x-axis? If so, how? If not, what changes, and why?

Q2 Write a conjecture in words for how the two area functions $\int_{-6}^{x_P} f(x)dx$ and $\int_{x_{start}}^{x_P} f(x)dx$ are related.

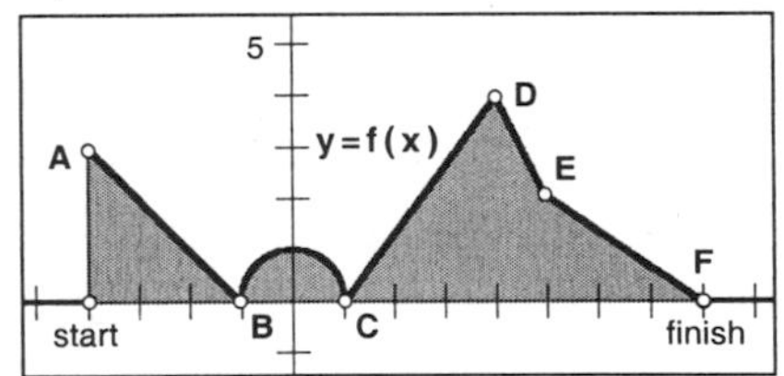

Be sure to keep points A, B, C, D, E, and F lined up in that order from left to right. If point C moves to the right of point D, the line segment CD will no longer exist.

2. Make a new shape for your area function by moving one or more points—point A, B, D, E, or F. Then move point *start* again along the x-axis.

Q3 Does your conjecture from Q2 still hold? Write the conjecture in integral notation.

3. Fix point *start* at the origin. Move point P to the left of the origin but to the right of point B.

Q4 The following two sentences sound good, but lead to a contradiction. Where is the error?

> The semicircle is above the x-axis from the origin to point P, so the area is positive. Point I, which plots the area, is below the x-axis, so the area is negative.

Area and Integrals

Name(s): ______________________

The shaded region in the figure at right is bounded by the x-axis, a function f, and two vertical lines, $x = 0$ and $x = 10$. You have seen that you can approximate the area of such a region with rectangles and trapezoids, and that this area is called the integral. In this activity, you will explore how limits are used to define the integral.

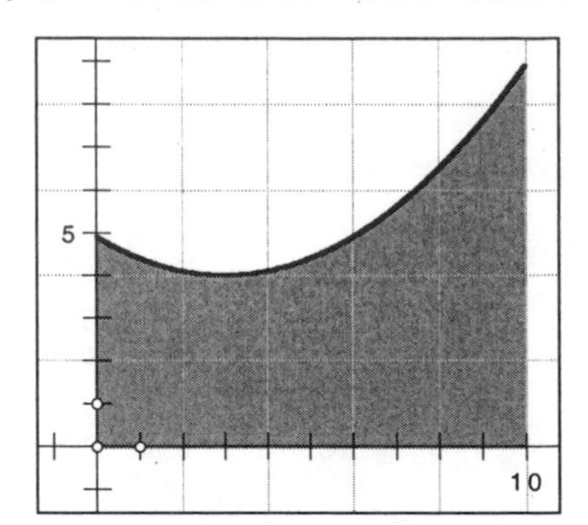

Sketch and Investigate

1. **Open** the document **AreaIntegral.gsp** in the **Exploring Integrals** folder. On page 1 you will find the plot of a function $f(x)$.

2. Press the *Show Integral* button to display the integral of f from point *start* to point *end*. The x-coordinate of point *start* is a, and the x-coordinate of point *end* is b.

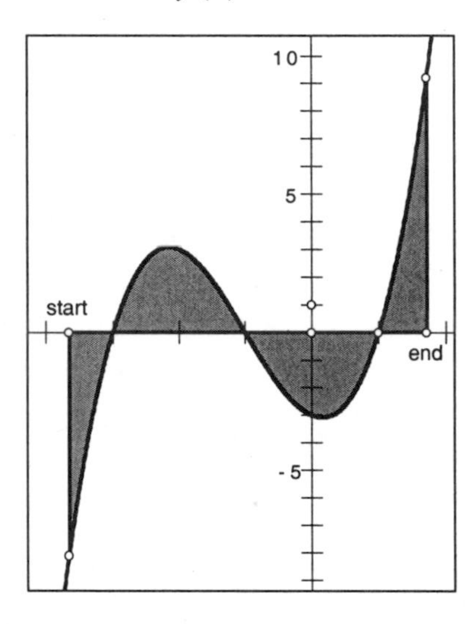

In the last activity you learned that the notation for this shaded region, or the integral, is $\int_a^b f$ and that in words we say "the integral from $x = a$ to $x = b$ of the function f." This is just the notation for the integral. The following steps will build the definition.

3. Go to page 2 of the document.

4. Choose **Riemann Rectangles** from **Custom** tools. This tool constructs a special type of rectangle that you can use to define the integral.

You'll actually be clicking on a segment which is on the x-axis.

5. To use this tool, click on point *start*, and then click on the x-axis a little to the right. As you move to make your second click, two points will be constructed. One marks the end of the new segment and a rectangle. The other point, called a *sample point*, is on the base of the rectangle and can be dragged. The tool uses the function's value at this point for the rectangle's height.

The label has been left off to avoid clutter.

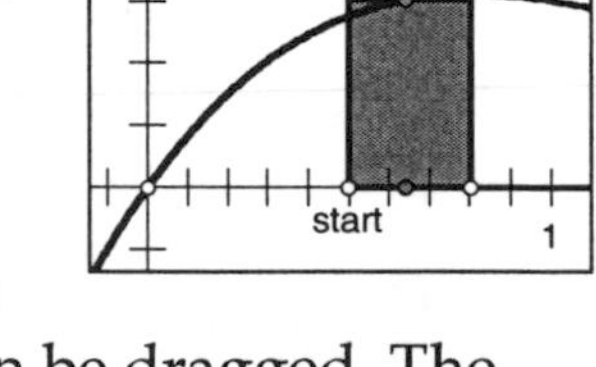

If you end up short of point *end*, drag the last constructed point to point *end* and adjust the others if you'd like.

6. Repeat step 5 four more times, using the right endpoint of the previous rectangle for your first point each time. Finish at point *end*. The widths of these rectangles do not need to be the same.

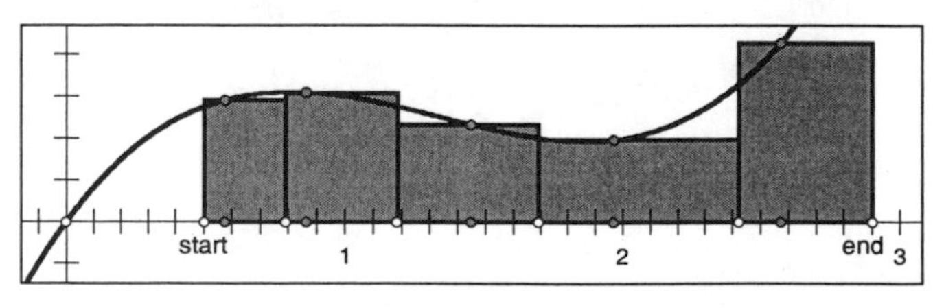

The tool also calculates the area of each rectangle—the width h times the height $f(x_i)$, where x_i is the sample point of the rectangle. This calculation is the building block of the integral's definition.

The rectangles you have made partition the interval $[a, b]$ into 5 subdivisions, each with area $f(x_i) \cdot h$ for $i = 1$ to 5. The sum of the areas of these rectangles is called a *Riemann sum*. If you place the sample point so that $f(x_i)$ is at the lowest y-value on the subinterval, the Riemann sum is called a *lower sum*.

Here, you will need to drag the light blue sample points on the x-axis, not the points on the function.

7. Position your sample points to create a lower sum by moving each one to the point where the function's value is lowest (see figure).

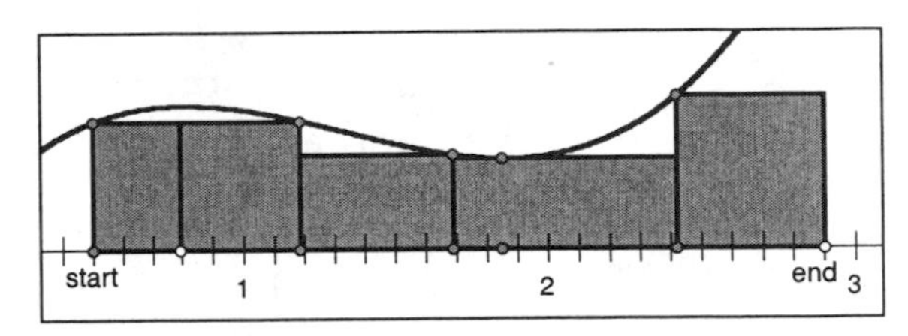

Click on a measurement to enter it into the calculator.

8. Choose **Calculate** from the Measure menu and calculate the lower sum by summing up all five product measurements $f(x_i) \cdot h$. Label this measurement *Sum* with the **Text** tool.

Q1 Write down the value you got in step 8 for the lower Riemann sum. Is it an overestimate or underestimate for the integral from $x = a$ to $x = b$?

An *upper sum* is a Riemann sum where the sample points are positioned so that $f(x_i)$ is the greatest y-value on the subinterval.

Check the status message to make sure you have point i. If you don't, click on that point again.

9. Position your sample points to create an upper sum by moving each sample point to where the function's value is greatest.

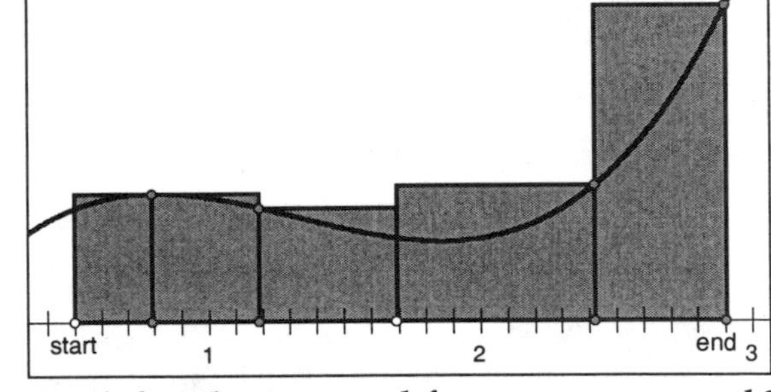

Q2 Write down the value for your upper Riemann sum. Is it an overestimate or underestimate for the integral from $x = a$ to $x = b$?

Q3 Why do you think there is a large difference between the lower sum and upper sum values that you found? What could you do to decrease this difference?

10. Go to page 3 of the document. On this page, there is the plot of a function $f(x)$ with rectangles already constructed to calculate two Riemann sums—one that underestimates and one that overestimates the actual area under the curve.

11. Adjust the slider for n so that there are six rectangles. Move the point on the *Under Over* slider so that the rectangles form a Riemann sum that overestimates the area. Then move the point so they form an underestimate for the area.

Q4 Record both Riemann sums for this example with $n = 6$.

12. Adjust the slider for n so that there are 12 rectangles and use the *Under Over* slider to record both Riemann sums for $n = 12$.

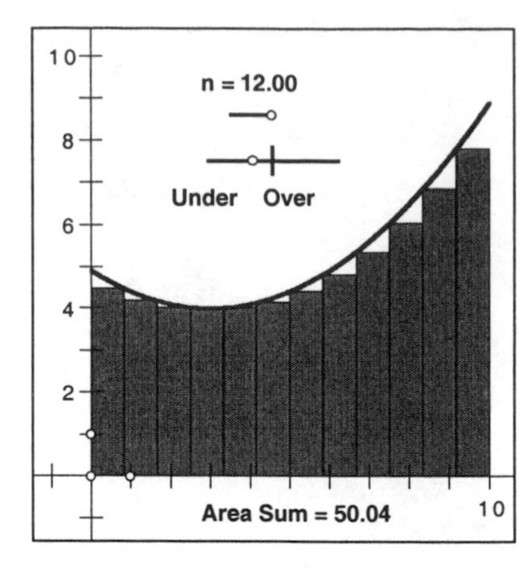

You can continue dragging the slider point outside the window, or expand your window to reach $n = 396$.

13. Repeat step 12 for $n = 24, 48, 96$, and 396. Make a chart of the Riemann sums for the different values of n.

If the limit of the Riemann sums exists as you make the largest of your h values approach 0, then the function is said to be *integrable* on the interval $[a, b]$, and this limit is called the *definite integral* of f on the interval $[a, b]$.

Try averaging the two Riemann sums for each n.

Q5 Does it look like the Riemann sums are converging to the same limit in step 13? If so, to what limit? If not, explain why not.

Q6 Check your answer by pressing the *Show Area* button.

So the integral $\int_a^b f(x)dx$ is defined as the limit of the sum of the n areas as h approaches 0 on the interval $[a, b]$, or $\int_a^b f(x)\,dx = \lim_{h_i \to 0} \sum f(x_i) \cdot h_i$.

In this example, n must be at least 2 to create the over-estimate. (Why?)

Q7 Go to page 4 and estimate $\int_1^{10} f(x)\,dx$ using the limit of the Riemann sums as you did in steps 11–13.

With this definition in hand, you can investigate some special properties of integrals.

14. Go to page 5 of the document. Shown is the constant function $f(x) = d$ where $d = 3$.

Q8 To find the value of $\int_{-2}^{3} f(x)\,dx$, you need only one Riemann rectangle. What is the value of this integral? Choose **Riemann Rectangles** from **Custom** tools to check. Click on point *start* and then point *end*.

15. Use the **Riemann Rectangles** tool again, but this time click on point *end* first and then point *start*.

Q9 What happens? What has changed in the calculation of $f(x_i) \cdot h$ to cause this to occur?

Q10 Write an integral that expresses the area that you just found in Q9.

16. Use the sliders to create the function $f(x) \approx -3$.

Q11 What is the value of $\int_{-2}^{3} f(x)\,dx$ for this function? $\int_3^{-2} f(x)\,dx$?

Now that you have done some examples and worked with Riemann rectangles, go back to the function on page 1. Without creating any rectangles, for each of the following integrals, decide whether the integral's value is a positive or negative number and explain how the definition of the integral helps you decide.

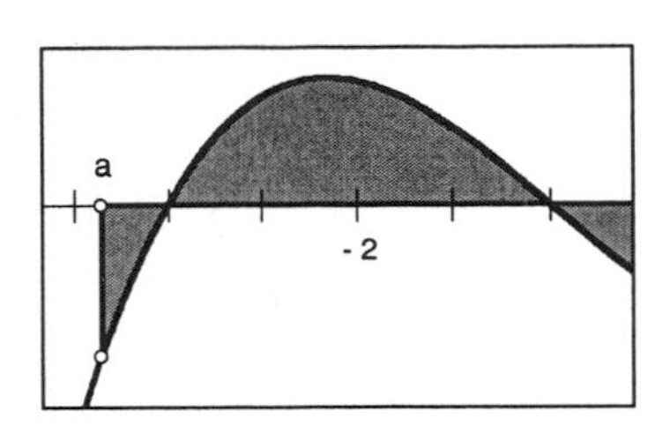

Q12 $\int_{-3}^{-1} f(x)\,dx$

Q13 $\int_{-1}^{1} f(x)\,dx$

Q14 $\int_{-3.5}^{-1} f(x)\,dx$

Q15 $\int_{-3}^{1} f(x)\,dx$ if the function $f(x)$ is symmetric around $x = -1$.

Q16 Using your answers above and what you now know about approximating integrals with rectangles and trapezoids, what approximation would you give for the integral of $f(x)$ from $x = -3.5$ to $x = 1.5$ for the function on page 1? Explain.

Explore More

Is it possible to add integrals? For any function f, does

$$\int_a^b f(x)\,dx + \int_b^c f(x)\,dx = \int_a^c f(x)\,dx?$$

Use the definition of the integral to explain your answer for each case.

Q1 If the value of $\int_1^5 f(x)\,dx = 9$, and $\int_1^5 g(x)\,dx = 3$, would $\int_1^5 (f(x) + g(x))\,dx = 12$? Explain, and sketch two functions that support your answer.

Q2 If the value of $\int_1^5 f(x)\,dx = 9$, would $\int_1^5 2f(x)dx = 18$? Explain and sketch a function that supports your answer.

The Area Function

Name(s): ______________________

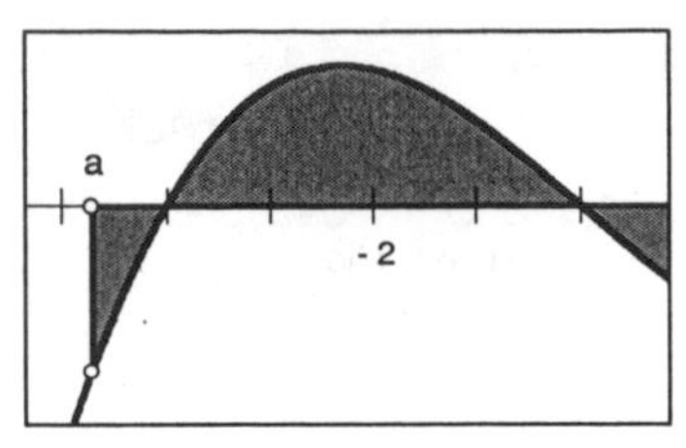

In previous activities, you used tools to approximate the area between a curve f and the x-axis from $x = a$ and $x = b$. By finding the limit of your approximations, you calculated the integral—the actual signed area—with a high degree of accuracy. But what if you wanted to find the signed area from $x = a$ to $x = b$ where $b = -4, -3, -2, \ldots$? Calculating all these limits, one at a time, would take forever. What we really need is a function that calculates the integral of f from $x = a$ to any point on the x-axis. Does such a function exist?

Sketch and Investigate

1. **Open** the document **AreaFunction.gsp** in the **Exploring Integrals** folder. On page 1 you will find the plot of a function $f(x)$, point *start* $(a, 0)$, and point *end* $(b, 0)$, where a and b are parameters.

Suppose that you want to find the value of the integral of f from $a = -4$ to any point $x = x_P$, or $\int_{-4}^{x_P} f(x)\,dx$. Does this expression define a function?

Double-click on the point with the **Text** tool to change its label.

2. With the **Point** tool, construct a point on the x-axis. Choose **Measure | Abscissa (x)** to measure the x-coordinate. Label the point P.
3. Choose **Measure | Calculate** and calculate the value of $f(x_P)$.

Select x_P and $f(x_P)$, in that order. Choose **Graph | Plot As (x, y)**.

4. Plot the point $(x_P, f(x_P))$.

In this case, the integral $\int_{-4}^{x_P} f(x)\,dx$ is equal to the area of the rectangle from $a = -4$ to $x = x_P$, with width $x_P - a$ and height $f(x_P)$.

5. Calculate this area by choosing **Calculate** from the Measure menu. Label this calculation *Area*(P) with your **Text** tool.

Select the four points in order: point start, point P, $(x_P, f(x_P))$, and $(a, f(a))$.

6. To illustrate the integral, plot the point $(a, f(a))$. Then select the four points that define the rectangle and choose **Construct | Quadrilateral Interior**.

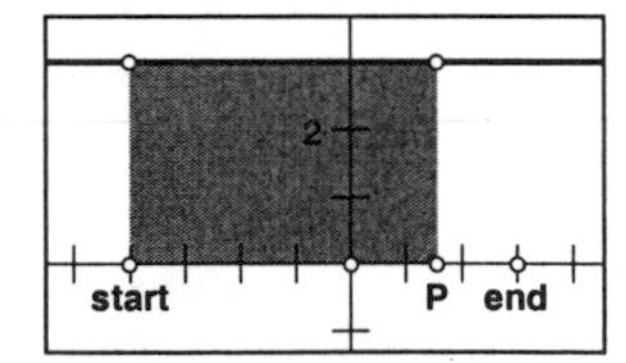

7. Move point P to different places on the x-axis—go left and right of point *start*. Observe the measurement *Area*(P). Can you tell whether this measurement is a function?

The measurement *Area*(P) is the integral of f from $x = a$ to $x = x_P$, but it's hard to tell whether it defines a function just by observing its values. A plot of the values for measurement *Area*(P) would help.

Remember to enter subscripts in brackets.

8. At each point P, where $x = x_P$, we want the y-coordinate to be the accumulated area under f from $a = -4$ to point P, or *Area*(P). So, plot

the point (x_P, *Area*(P)). Label this new point A_P. (If you can't see the new point, drag point P closer to point *start*.)

To make a smoother trace, choose **Display | Animate Point**. Then decrease the speed on the Motion Controller.

9. With point A_P selected, choose **Display | Trace Plotted Point** to turn on tracing for point A_P. Slowly drag point P along the x-axis to the left and right of point *start*.

Q1 This trace is the plot of $A(x_P) = \int_{-4}^{x_P} f(x)\,dx$. Does the trace look like it defines a function? (Does it pass the vertical line test?) If you move point P to the left of $a = -4$, the rectangular area is above the x-axis, but *Area*(P) is negative. Why?

Q2 Where is $A(x_P)$ zero? Increasing? Decreasing?

Q3 What type of plot is the trace itself? Write an expression for the function that you think the point is tracing. Can you prove that your answer is correct?

10. Go to page 2 to examine $A(x_P) = \int_{a}^{x_P} f(x)\,dx$ for $a = -4$ and a different $f(x)$. Here, f is linear, so the region between the plot of f and the x-axis is not a rectangle, and you will need to calculate the area differently.

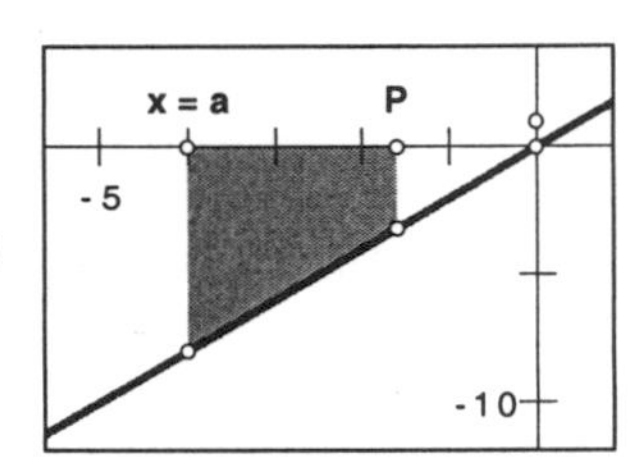

See the Activity Notes for an explanation of why you can use the area formula for a trapezoid here.

11. Use the formula for the area of a trapezoid to calculate the area from $a = -4$ to $x = x_P$. Label this calculation *Area*(P).

12. As before, plot the point (x_P, *Area*(P)). Turn on tracing for this point.

Reducing the speed of the animation will trace in more points.

13. With just point P selected, choose **Display | Animate Point** to move point P along the x-axis and to trace the values of the integral $\int_{-4}^{x_P} f(x)\,dx$ as x_P varies.

Q4 Where is $A(x_P) = \int_{-4}^{x_P} f(x)\,dx = 0$? Where is it increasing? Decreasing?

Q5 Describe what happens to the plot of $A(x_P)$ where $f(x) = 0$.

Q6 What type of curve is the trace itself? Write an expression for the function that you think the point is tracing. How can you prove that your answer is correct? (*Hint:* Use the function $f(x) = 2x$ in the formula for the area of the trapezoid.)

Q7 Pick any constant. If you substitute this constant into your expression from Q6, does it give the same value as the area from $a = -4$ to your constant?

In the cases above, you can find an expression that tells you the value of $A(x_P)$ for any value of x_P. To evaluate $\int_{-4}^{b} f(x)dx$, all you need to do is substitute b into this expression. For functions that are not linear or constant, you cannot use a formula from geometry to create the expression, but you can use limits.

You have built rectangles and calculated the sum of their areas to estimate the integral before. Here, you'll build one rectangle, and use it to trace an approximation of the plot of $A(x_P) = \int_{a}^{x_P} f(x)\,dx$, for $a = -4$.

14. Go to page 3. On this page, you will find a cubic function $f(x)$ and an independent point P. You will use point P to trace the plot of the approximate integral function.

Remember to enter subscripts in brackets.

15. Use the parameter *zero* to plot the point $(x_P, 0)$. Label this point P_0.
16. Calculate $f(x_P)$ and plot the point $(x_P, f(x_P))$.

Again, we need to make a small rectangle so the estimate is fairly accurate. The small adjustable measurement h is used for the width of our rectangle.

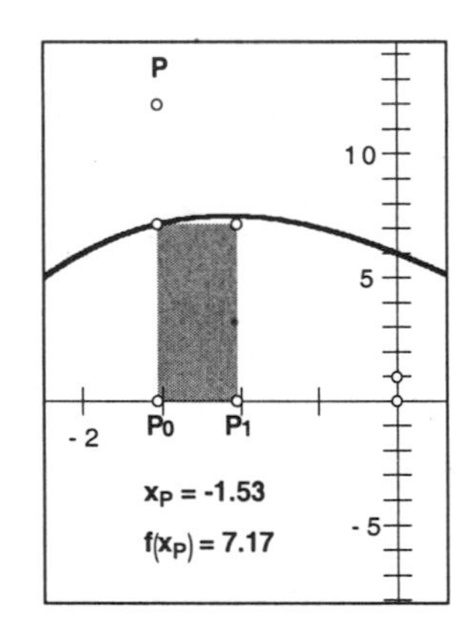

17. To make the base of the rectangle, calculate $x_P + h$. Plot the point $(x_P + h, 0)$. Label this point P_1.
18. We want the right side of the rectangle to be the same height as the left side, which is $f(x_P)$. So plot the point $(x_P + h, f(x_P))$. Now you have four points that form a rectangle.

Select the points in clockwise or counterclockwise order.

19. Construct the interior of the rectangle by selecting the four new points and choosing **Construct | Quadrilateral Interior**.
20. Finally, calculate the area of this rectangle, $h \cdot f(x_P)$. Label this calculation *Area*(P).

Now, to sum up, or accumulate, the *Area*(P) values as x_P varies, we need a starting value. We will use the y-coordinate of point P.

To see the accumulation, construct a segment from point P to point Q.

21. From point P_0 to P_1, only the area of the rectangle, *Area*(P), will be accumulated, so calculate the value y_P + *Area*(P). Plot the point ($x_P + h$, y_P + *Area*(P)) and label this point Q.

Now, you can plot the approximate value of the integral function, $A(x_P) = \int_{a}^{x_P} f(x)\,dx$ where $a = -4$, by using points P and Q.

22. Because y_P is the starting value for accumulating the area, if we move point P so that its x-coordinate is $x_P = -4$ then y_P must be 0. (Why?) So press the *Move P→a* button.

Check the status line to make sure you have selected point *P*. If not, click on the point again.

23. The *y*-value of point *Q* is now the approximate value of $A(x) = \int_a^x f$ at $x = a + h$. To trace the value of this integral as *x* varies, select point *P*, then point *Q*, and choose **Edit | Action Buttons | Movement**. Choose **instant** speed and click OK.

By tracing the rectangle, you can see if your trace is underestimating or overestimating the integral.

24. Select point *P* and the rectangle, then choose **Display | Trace Objects** to turn on tracing.

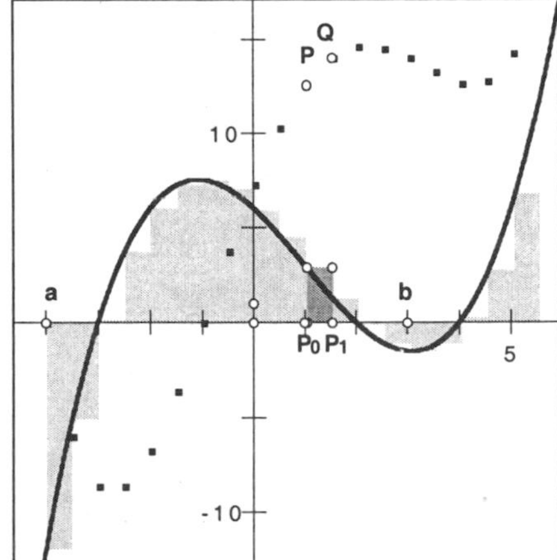

25. Press the *Move P→Q* button a few times with $h \approx 0.5$. (If *h* is negative, the movement will be to the left.)

Q8 Can you tell whether your trace is an overestimate or an underestimate?

Select the *Move* button and choose **Edit | Properties**. On the Move panel, change speed to **instant**. Click OK.

26. To see a more accurate trace, adjust the slider for *h* to a smaller value, and your *Move P→Q* button to **medium** speed. Press the *Move P→a* button and then press the *Move P→Q* button.

Although this plot is only an approximation of the function $A(x_P) = \int_a^{x_P} f(x)\,dx$, it gives a good idea of the plot of the function. In the next two activities, you will explore how to find an expression for the integral function when you can't use a geometric formula.

Explore More

You can edit the expression for $f(x)$ or the value of the parameter *a* to create traces of $A(x_P) = \int_a^{x_P} f(x)\,dx$ for other functions and values of *a*.

Go to page 4 and experiment with tracing the integral approximation for values less than $x = a$ by adjusting the *Direction* slider to –1 and pressing the *Move P→a* button. Try tracing the integral function on both sides of *a* for each of the combinations below.

Q1 Try $f(x) = 1/x$ for $a = 1$. Be careful as point *P* approaches the discontinuity in the graph of *f*. Do you recognize this function?

Q2 Try $f(x) = x^2$ for $a = 0$. Do you recognize this function?

Q3 Try $f(x) = e^x$ for $a = 0$, as well. What is this function?

Plotting the Integral

Name(s): ______________________

In the previous activity, you explored the concept of an *area function* by plotting the accumulated signed area from a fixed point ($a = -4$) to a varying value x_P.

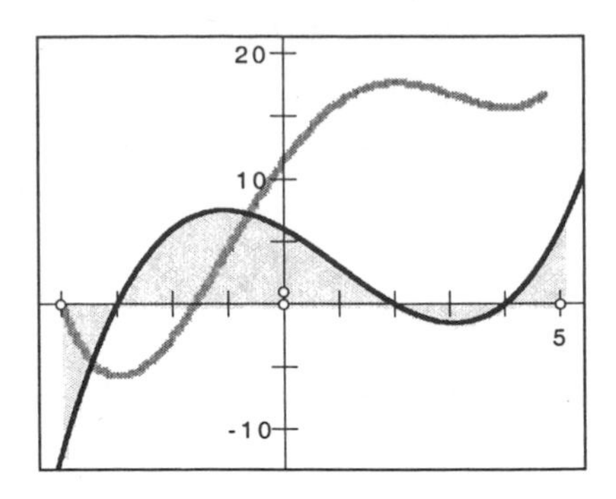

In this activity, you will investigate the properties of functions *defined* by integrals. Along the way, you will explore how to find expressions for these functions and learn how to sketch them using paper and pencil.

Sketch and Investigate

1. **Open** the document **PlotIntegral.gsp** in the **Exploring Integrals** folder. On page 1 of this document you will find the plot of a function f and some measurements.

To evaluate $\int_a^{x_P} f$ for any values of a and x_P, it would be convenient to have an expression into which you could substitute values directly. Coming up with such an expression that works for all values of a and x_P requires some experimenting.

All of these measurements and objects were constructed in the previous activity.

2. If f is constant, then $\int_a^{x_P} f$ is equal to the area, labeled *Area*(P), of the rectangle from $x = a$ to $x = x_P$. Point A_P has coordinates (x_P, *Area*(P)). Move point P to see the trace of the integral function for the given a.

3. To create a dynamic plot of the integral function for different values of a, select points A_P and P. Choose **Locus** from the Construct menu.

To change the value of a, move the point on the x-axis where $x = a$.

4. You now have a plot of $A(x_P) = \int_a^{x_P} f$ where both a and x_P can vary. Experiment with moving point P, then with changing the value of a.

Q1 For any value of a, where is the integral function zero? Increasing? Decreasing?

Q2 What happens to the locus when you move point P? When you change the value of a?

Q3 Write an expression for this locus for all values of a. (Your expression will involve x_P and a.)

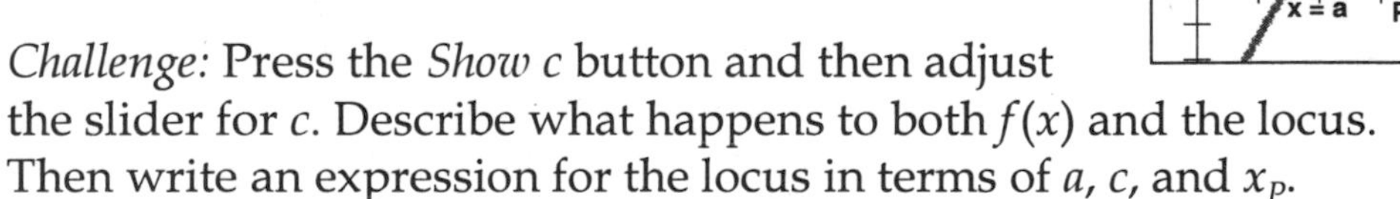

Q4 *Challenge:* Press the *Show c* button and then adjust the slider for c. Describe what happens to both $f(x)$ and the locus. Then write an expression for the locus in terms of a, c, and x_P.

Now let's see what happens for a different function.

Plotting the Integral (continued)

See the previous activity to learn how the objects here were constructed. The trapezoid formula was used for *Area*(*P*) here.

5. Go to page 2 of the document. Here f is a linear function. Again, point A_P plots the point $(x_P, Area(P))$ where $Area(P)$ is the value of the integral from $x = a$ to $x = x_P$, and both a and x_P can vary.

6. Select points A_P and P, then choose **Locus** from the Construct menu.

7. You now have a plot of $A(x_P) = \int_a^{x_P} f$ where both a and x_P can vary. Experiment with moving point P, then with changing the value of a.

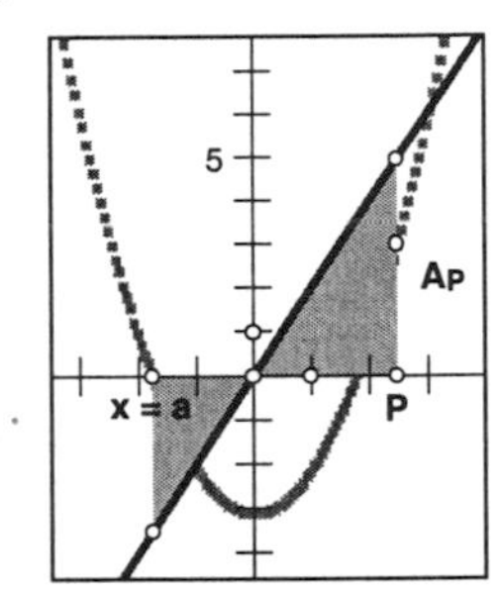

Q5 For any value of a, where is the integral function zero? Increasing? Decreasing?

Q6 What happens in your sketch when you move point P? When you change the value of a?

Q7 Write an expression for this locus for all values of a. (Your expression will involve x_P and a.)

Another challenge: Can you answer Q4 for this function?

Q8 *Challenge:* In both examples above, varying a resulted in a family of vertically translated functions. Can you determine the distance between the plots for two different values of a?

Once you go beyond linear functions, you can't use a geometric formula for $Area(P)$. Instead, we will use approximations and limits.

See the previous activity for the construction steps or for more explanation.

8. Go to page 3. Here you will find a plot of $f(x) = \sin(x)$, points P and Q, and a trapezoid. The trapezoid approximates the integral of f on the interval $x = a$ to $x = a + h$. You will use these objects to construct a dynamic approximation of $A(x) = \int_a^x f$ using iteration.

9. The difference between the y-coordinates of points P and Q is the signed area of the trapezoid. Move point P and track how point Q moves in response to the changes in the trapezoid and its area.

As in the previous activity, the y-coordinate of point P will be the starting point for accumulating the area.

10. Because $A(x) = \int_a^x f = 0$ at $x = a$, merge point P with the point at $x = a$. To do this, select point P and then the point where $x = a$. Choose **Merge Points** from the Edit menu.

Point P is now the point where $x = a$.

11. Select the point where $x = a$ and choose **Transform | Iterate** and then click on point Q for the First Image. Uncheck Tabulate Iterated Values from the Structure pop-up menu, then press **Iterate.** The images of three more trapezoids will be constructed. Also, the image of point Q will create an approximate plot of $A(x)$.

To select an iterated image, click once on any of the newly constructed images.

12. Select the iterated image of the trapezoids and press the plus (+) key to increase the number of iterations and see more of the plot.

Q9 For this approximate plot of $A(x) = \int_a^x f$, where is $A(x) = 0$? Increasing? Decreasing?

Q10 What will the plot of $A(x)$ look like for values of x less than a? Sketch your prediction in the margin.

Q11 Go to page 4 to see the answer to Q10. Move around the point where $x = a$. Describe the family of curves for $A(x)$, for varying a.

A basic function is one that has no transformations— x and $\sin(x)$ are basic functions; $x + 3$, $5x$, $\sin(3x)$, and $\sin(x) + 4$ are not.

Q12 Does the plot of $A(x)$ look familiar? Write down which basic function you think is being approximated here.

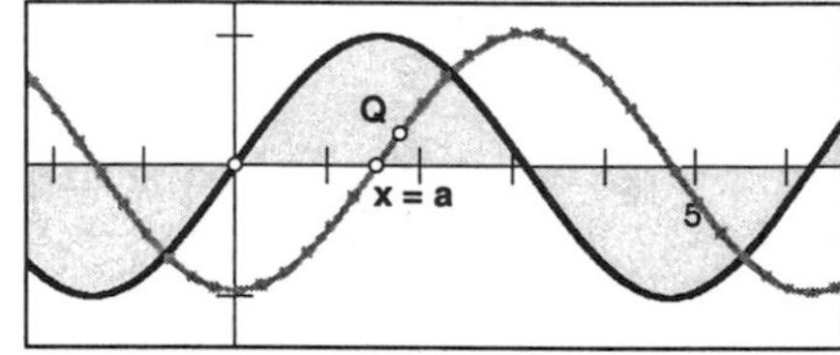

13. Choose **Graph | Plot New Function**. Enter your prediction from Q12. Does your plot match the iterated image's shape for all values of a (ignoring translations)?

Q13 *Challenge:* Now try to match the plot exactly, including the translation. Double-click on your prediction to edit the expression. Can you determine the role of a in the expression that will match the plot?

You have looked at three different types of functions and predicted very specific expressions for the integral function $A(x) = \int_a^x f$. Now let's look for some general relationships. Look back over pages 1–4 and use page 5, as well, to answer these questions and sum up the relationship between f and $A(x)$.

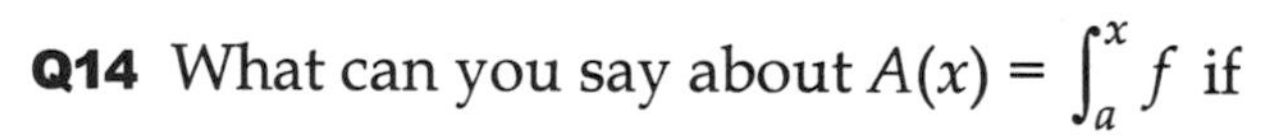

Q14 What can you say about $A(x) = \int_a^x f$ if

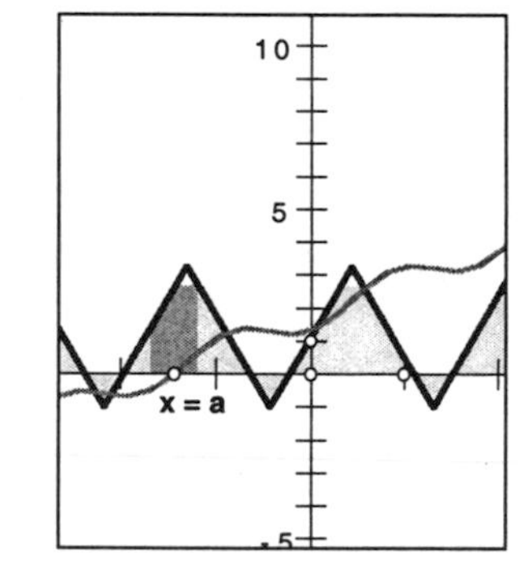

a. f is positive? f is negative?

b. f changes sign?

c. f is increasing?

d. f is decreasing?

e. f has a maximum or minimum?

Q15 What can you say about f when

a. $A(x)$ is increasing?

b. $A(x)$ is decreasing?

c. $A(x)$ has a maximum or minimum?

d. $A(x)$ is translated vertically?

Plotting the Integral (continued)

Are the connections between the plots of f and $A(x)$ familiar? Make a conjecture about the relationship between f and $A(x)$ based on your answers to Q14 and Q15.

Explore More

Earlier, you constructed an approximate plot of the integral using iteration. Here, to gain more experience sketching a plot of the integral function by hand, you will use a tool that mimics how you might draw it yourself with paper and pencil.

Go back to steps 8 and 9 if you have forgotten what these objects represent.

1. Go to page 6. Choose **Plot Area with Trapezoids** from **Custom** tools. With the tool click once on the point where $x = a$ to create a trapezoid and a segment.

This tool will automatically match the function f and the parameter h and create a trapezoid between the function plot and the x-axis. It will also construct two points whose y-values differ by the signed area of the trapezoid and the segment between these two points.

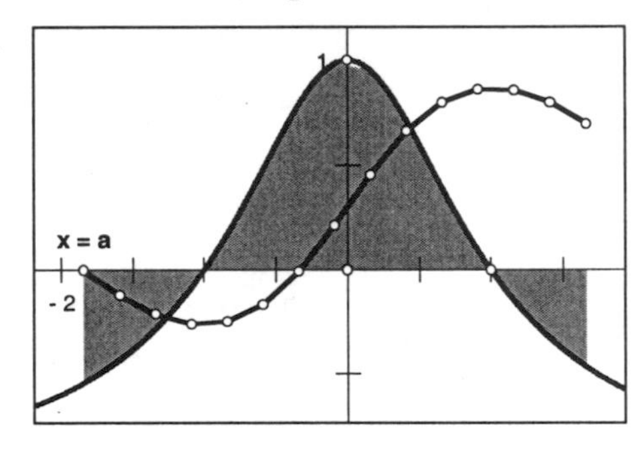

2. Using the tool again, click on the right endpoint of the segment constructed in step 1. This will add the area of the next trapezoid to the first one.
3. Continue using the tool to build the approximate plot of the integral function, each time using the right endpoint of the previously constructed segment.
4. Choose **Plot Area with Trapezoids(left)** from **Custom** tools. Click again on the point where $x = a$ to extend your plot to the left. Continue using this tool, each time clicking on the left endpoint of the previous segment.

Q1 When will the constructed segment have a positive slope? A negative slope?

Q2 How can you tell whether the next constructed segment will be sloped at a greater or lesser angle than the previous segment?

Q3 When will the plot you make have a maximum? A minimum?

Go to page 7 and sketch the plot of $A(x) = \int_a^x f$ for the given function in the margin, using your answers from Q1–Q3 above. Then check your answer using the tool, step by step. How did you do?

Getting Down to Fundamentals

Name(s): ______________________

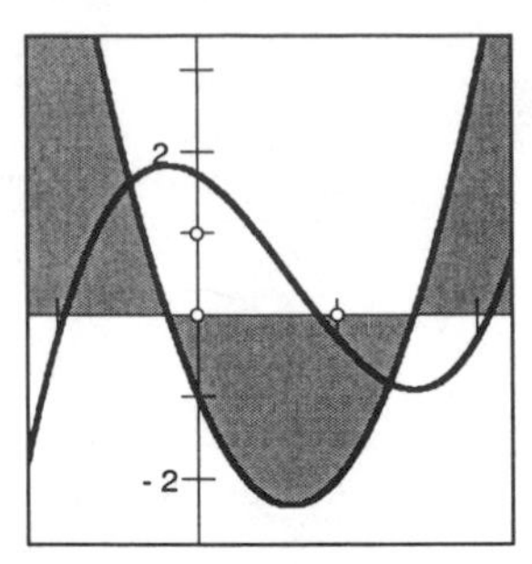

In the last activity, you explored properties of the function $A(x) = \int_a^x f$ for any a. The relationships that you found between the plots of f and $A(x)$ hinted that $A(x)$ is an antiderivative of $f(x)$. In this activity, you will examine this possibility in a variety of ways.

Sketch and Investigate

First, let's see if a plot of $A(x) = \int_a^x f$ matches a symbolic antiderivative.

1. **Open** the document **Fundamentals.gsp** in the **Exploring Integrals** folder. On page 1 of this document you will find the plot of the function $f(x) = 3x^2 - 4x - 1$ and some sliders.

Click on measurement C to enter it into the function editor.

2. Choose **Plot New Function** from the Graph menu, and enter the antiderivative of $f(x)$, $x^3 - 2x^2 - x + C$.

Leave the plot's thickness as **Thin**.

3. Double-click on the equation for the antiderivative with your **Text** tool and label it F.
4. Press the $C \rightarrow 2$ button so that f has an intercept at $x = -1$.

Now you will use a tool from a previous activity to trace a plot of the integral function $A(x)$ and see if the trace is identical to the plot of $F(x)$.

5. Choose **Area with Trapezoids** from **Custom** tools. This tool will automatically match the function $f(x)$ and the parameter h.
6. With the tool, click in an empty spot in the plane. With the **Text** tool, label this point P. Label the other point Q.

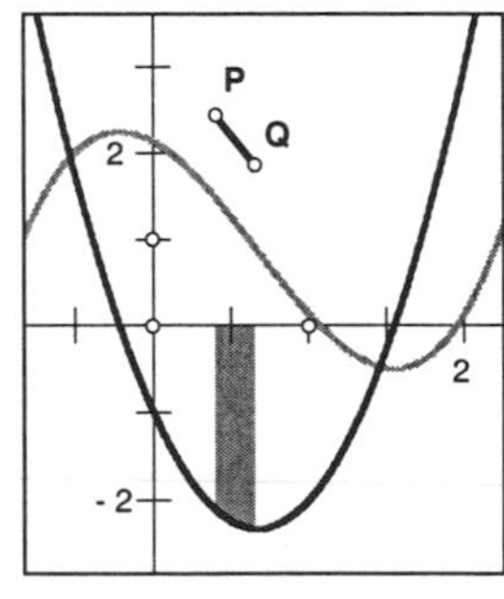

This tool also creates a trapezoid whose area approximates the integral of $f(x)$ on $[x_P, x_Q]$. Now, you can create an approximated trace of $A(x)$ with these two points by having point P *chase* point Q.

Select point P, then point Q, and choose **Edit | Action Buttons | Movement.**

7. Make a button that will move point P to point Q at **medium** speed.
8. Move point P to $(-1, 0)$. Then select point P and the trapezoid and choose **Trace Objects** from the Display menu.
9. Press the $h \rightarrow 0.01$ button so you will get a very close approximation of $A(x)$ when you create the trace.

If point P moves off your screen, choose **Edit | Undo Animate Point.**

10. Then press the *Move* $P \rightarrow Q$ button to create an approximate trace of the integral function.

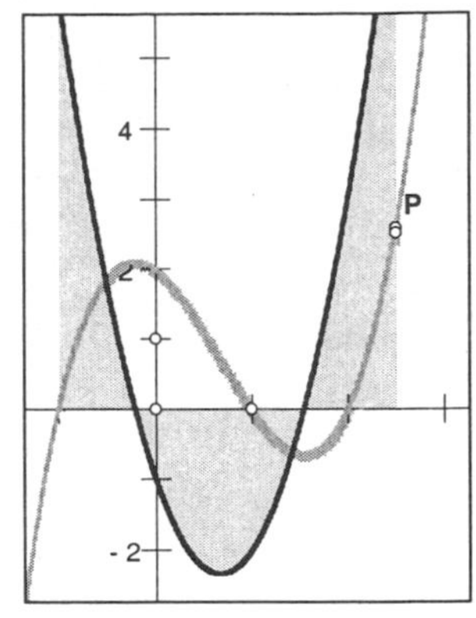

Your trace of $A(x) = \int_a^x f$ and the plot of $F(x)$ should be almost identical. Try this again. Adjust the slider for C to any value, then move point P to any point on the plot of F, erase your traces, and then press the *Move P→Q* button. Does it work again?

To erase a trace, press the Esc key twice.

Is $A(x) = \int_a^x f$ really an antiderivative of f? If it is, then $A'(x)$ must be $f(x)$. Let's see if this is the case.

See the previous activity to learn how to construct $A(x)$ using iteration.

11. Go to page 2. Here you will find a plot of an approximation of $A(x)$ constructed using iteration. Although this plot is only an approximation, you can still use it to look for evidence that its derivative is $f(x)$. We'll test this idea graphically by plotting the slope of some secants to the plot of $A(x)$, and then taking a limit.

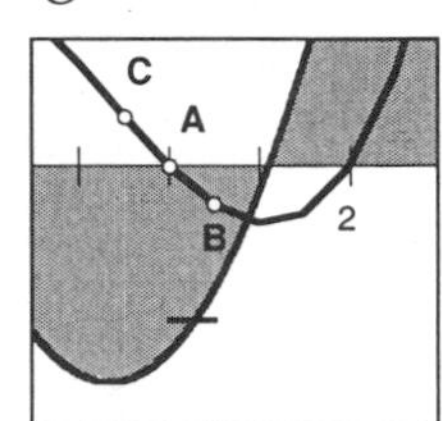

12. Choose **Secant Slope** from **Custom** tools. With the tool, click on points A and B. This tool creates a horizontal segment—a "step"—with y-value equal to slope AB, on the interval $[x_A, x_B]$.

Press the *Hide Area* button before performing the iteration to avoid creating unnecessary objects.

13. With the **Arrow** tool, select point A for the pre-image point and choose **Transform | Iterate**. For the first image, click on point B. Uncheck **Tabulate Iterated Values** in the Structure pop-up menu. Press the plus (+) key until **Number of Iterations** is 10. Click Iterate.

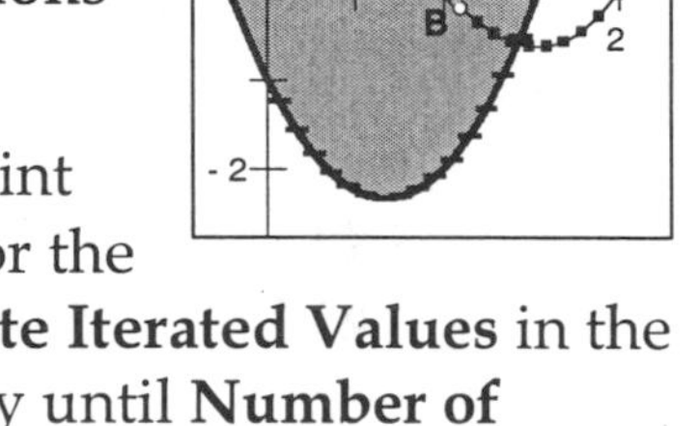

14. Select just point A again for the pre-image point and choose **Transform | Iterate**. This time, for the first image, click on point C. Uncheck **Tabulate Iterated Values** in the Structure pop-up menu. Press the plus (+) key until **Number of Iterations** is 10. Click Iterate.

15. Adjust the slider for h to a small value to approximate the limit.

Q1 Explain what happened graphically in step 15 at point A and to the "steps." How does this add evidence that $A'(x) = f(x)$ at point A?

Q2 If $A'(x) = f(x)$, then what is the value of $\lim_{x_B \to x_A} \frac{y_B - y_A}{x_B - x_A}$?

16. Test your answer to Q2. Make $h \approx 0.25$. Select segment AB and choose **Measure | Slope.** Then use the slider to adjust h and make point B approach point A. How did you do?

From the steps above, it looks like the slope of $A(x)$ is the y-value of $f(x)$ at point A. In fact, each iterated "step" intersected the plot of f. To prove that

$A'(x) = f(x)$ algebraically, we can apply the definition of the derivative to the function A.

Now the question is: Does $\lim_{h \to 0} \frac{A(x+h) - A(x)}{h}$ equal $f(x)$? You can answer this question by looking at this limit at an arbitrary point.

17. Go to page 3. Here you will see the integral of f from $x = a$ to $x = x_P$ shaded in light green. This area is equal to $A(x_P)$.

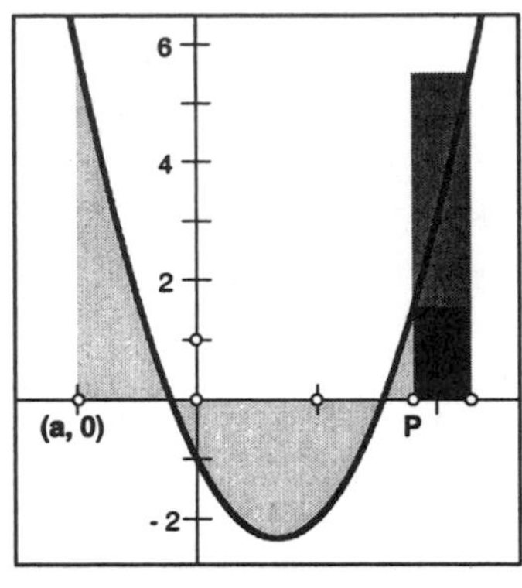

Q3 Press the *Show Locus* button to see the integral from $x = x_P$ to $x = x_P + h$. Explain why this integral equals $A(x_P + h) - A(x_P)$.

You can move point P to see how the areas change.

18. Hide this locus and then press the *Show Rectangles* button. One of these rectangles is an overestimate of the value of $A(x_P + h) - A(x_P)$, and the other is an underestimate.

Q4 Write an expression for the left rectangle in terms of f. (*Hint:* What is the rectangle's height? Width?)

Q5 Use the expression from Q4 in place of the numerator of the above limit to show that it equals $f(x_P)$.

Explore More

Above, you showed that $A(x)$ is an antiderivative of f. Here you'll look at this fundamental connection between slope and area in another way.

To plot F', select the equation for F' and choose **Graph | Plot Function**.

1. Go to page 4 of the document. Select the function F and choose **Graph | Derivative**. Plot $F'(x)$. Relabel the function F' as f.

You won't see anything created at first.

2. Choose **Riemann Area** from **Custom** tools. With this tool, click on the point $(a, 0)$ to start the area plot.
3. Click on the point $(a, 0)$ a second time to start the first rectangle and then click again on the x-axis, but a little to the right of the point $(a, 0)$. As you make this third point, the tool creates a random point between them, called a *sample point*. The tool uses the y-value of $f(x)$ at the sample point for the height of the rectangle and will also create a segment where the difference between the y-coordinates of the endpoints represents the signed area of the rectangle.

After you use the tool once, you only need to select the two points on the x-axis that determine the base of the rectangle. The tool automatically strings together the segments for the plot of accumulated area.

4. Use the tool to make seven more rectangles. Each time, match the first point on the x-axis with the last point of the previous interval. Your intervals do not need to be the same size.

The plot of approximated area probably does not match the plot of $F(x)$ very well, but you can apply a theorem about derivatives to make a better match.

5. First, move the left endpoint of your first rectangle to the x-intercept of $F(x)$ as in the figure at right.
6. Choose **Average Rate** from **Custom** tools. This tool uses points on the x-axis instead of points on the plot and automatically identifies $F(x)$.

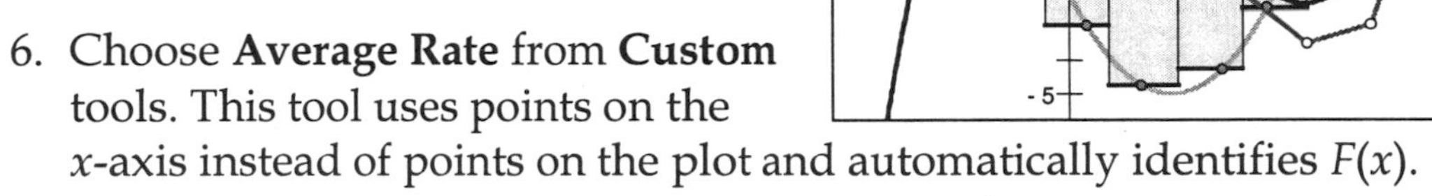

You will need to click twice on each interior point. Once to end a segment and once to start the next segment.

7. With the tool, click on the endpoints of each of the intervals you made above. This tool creates horizontal segments, each with y-value equal to the average rate of change of F on its interval.

Q1 What theorem about derivatives is demonstrated at the points where the horizontal segments (the average rates of change of F on the interval) intersect the plot of the derivative? Write out the theorem in its symbolic form.

Now, we'll use these intersections to place the sample points.

Or move the sample point so that the top of the rectangle and the horizontal step are the same.

8. Move the sample point in each interval so that its corresponding point on the rectangle's top is the intersection of the derivative plot and the horizontal segment.

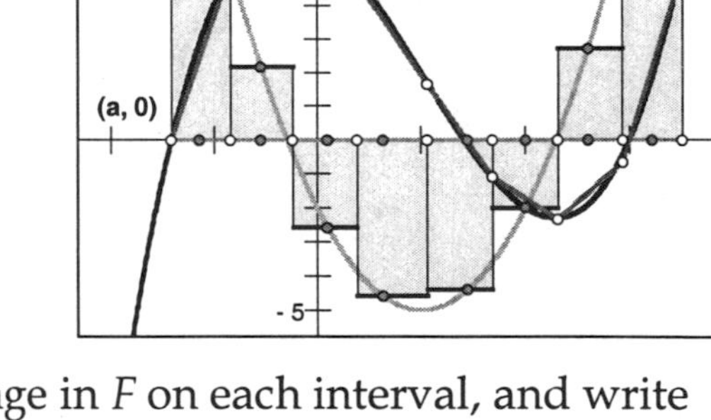

Q2 Use your answer to Q1 to explain why the area of each rectangle is now the *exact* value of the net change in F on each interval, and write an expression for the area of each rectangle in terms of F.

Q3 Explain why the sum of an arbitrary number of rectangles constructed in this way on an interval $[a, b]$ will equal $F(b) - F(a)$. (You can also use limits to prove that the integral of F' on $[a, b]$ equals $F(b) - F(a)$.)

Q4 Explain how to use the ideas in this activity to find the exact value of the integral of a function whose antiderivative you know.

Q5 Given a function f with $f(4) = -1.3$ and the function $g = \int_7^x f$, explain in your own words why $g'(4) = -1.3$.

Activity Notes

Visualizing Change: Position (page 3)

Prerequisites: Students should be familiar with the idea of plotting position as a function of time.

Sketchpad Proficiency: Beginner/Intermediate. Students will need to measure coordinates and move objects in the plane.

Activity Time: 45–55 minutes

Document: Position.gsp

General Notes: "A particle moves on the x-axis ..." begins many a calculus question. In this activity, your students will use a point on a segment to represent that "particle" and control its motion in order to visualize motion along a line and plots in the x-y plane arising from that motion.

In the Extension, you will see how to model the motion of a "particle," allowing you to create your own visualizations of such questions.

Sketch and Investigate

In the document **Position.gsp**, a point (labeled *Me*) has been constructed on a segment. The difference between the x-coordinate of this point and the x-coordinate of a fixed point (labeled *Home*) are used to create a measurement of the position of the particle *Me*. In **Q2**, when point *Me* is to the right of point *Home*, the position will be positive, and to the left, negative.

To simulate the passage of time, students will animate another point labeled *Time*. Using the x-coordinate of this point as *time* and the difference between the x-coordinates of the points *Me* and *Home* as *position*, students will plot (*time, position*) as a point on the coordinate grid. (This point is labeled P in the activity.)

What is created is essentially a "motion detector." As the point *Time* moves, the x-coordinate of point P will change. The y-coordinate of point P will only change if the student moves point *Me*. By tracing the location of point P, the student can visualize the motion of the particle with a coordinate graph.

In **Q3**, for example, when the *Animate* button is pressed, point *Time* will move, and if the student does not move point *Me* then the plotted point P will trace a horizontal line.

In **Q4**, moving point *Me* to the right will cause the y-coordinate of point P to increase, creating a trace with a positive slope. This occurs because the measurement of position is the difference in two x-coordinates, $x_{Me} - x_{Home}$. Moving point *Me* to the right increases this value. This is an important observation, because though a movement toward point *Home* from the left is a decrease in distance, it is an increase in relative position. In **Q5**, moving point *Me* to the left causes a decrease in relative position and a trace with a negative slope. You may wish to use this idea to begin a discussion of velocity.

Explore More

A control allows students to adjust the velocity of a "particle" rather than the position. This is accomplished by having point *Me* "chase" a hidden point whose distance from *Me* is determined by the *velocity* slider.

Extension

To make your own motion detector, construct a point A and translate it 1 cm. Construct a ray from A through A'. Construct point B on this ray and measure the ratio AB/AA'. Relabel this ratio *time*.

Construct another line through two new points, C and D. Construct a point E on this line. Measure the coordinates of points C and E. Calculate the value of $x_E - x_C$ and label this calculation *position*. Plot the point (*time, position*). Create a button to animate point B forward, slowly.

To model a question such as "a particle moves so that its position at any time $t \geq 0$ is given by $x(t)$," you can build a point so that its position is determined by a function, rather than by movement of the mouse.

Follow the steps above to create a *time* measurement. Create a function $f(x)$ and calculate $f(time)$. Add $f(time)$ to x_C. Measure the y-coordinate of point C and then plot the point $(y_C, f(time))$.This point will have a location on the line CD determined by the value of the function you created. (Rather than being free to move along the line, as point *Me* was in the activity, you must change the value of *time* to have the point change position.) You can also create an *Animate* button as in the activity to create a position plot.

What does a particle that "moves along the x-axis so that its position at any time $t \geq 0$ is given by $x(t) = 2te^{-t}$" look like? (This question is from the 1993 AP test. Instead of line CD, construct point E on the x-axis. The origin will serve as point C.)

Visualizing Change: Velocity (page 7)

Prerequisites: Students should be familiar with the concept of velocity, as well as how to find the slope and equation of a line on a position vs. time plot.

Sketchpad Proficiency: Beginner/Intermediate. Students will need to measure coordinates and move objects in the plane.

Activity Time: 45–55 minutes

Document: Velocity.gsp

General Notes: In this activity, students will be able to visualize the sentence "A particle moves with velocity..." by controlling the velocity themselves, as you might do with a remote control toy car (or your own automobile, for that matter!). Students will observe both the motion of the particle—a point—on a line, as well as a plot of time and position. In the Extension, you will see how to create such a demonstration yourself.

Sketch and Investigate

In the document **Velocity.gsp**, a point *Me2d* has been constructed in the plane. The y-coordinate of this point has been used to create a point, *Me,* on a line at the top of the sketch. In **Q1**, the position of this point on the line is determined by the y-coordinate of point *Me2d*. If point *Me2d* has a y-coordinate of 3, then point *Me* will be 3 units from point *Home* (as determined by the scale of the y-axis). In **Q2**, if the y-coordinate of point *Me2d* does not change, neither will the location of point *Me*.

The x-coordinate of point *Me2d* represents time. When students show the velocity control, a point *FutureMe* will also be shown. This point shows the time and position of point *Me2d* if the point has constant velocity, as determined by the slider for velocity. In **Q3**, with the time control set to 0.5, and the velocity control set to 2, point *Me2d* will move right 0.5 unit and up 1 unit.

In **Q4**, different time settings will simply move point *FutureMe* along a line with a slope equal to the velocity setting. In **Q5**, changing the velocity setting will change the slope of the line (not shown) between point *Me2d* and point *FutureMe*. At all times, the slope of this line will equal the velocity setting.

In Sketchpad, you can make points "chase" each other, and that is what will happen when you press the *Animate Time* button. Point *Me2d* will move toward point *FutureMe*. In **Q6** through **Q10**, you will see point *Me2d* move along a line with a slope equal to the velocity setting. With the velocity set to 0, the point will create a horizontal line, and point *Me* will not move at all.

Even more interesting is what happens when you change the velocity during the "chase." Point *Me2d* will try to follow the point *FutureMe* at all times, even as point *FutureMe* moves itself. By changing the velocity value, you change the slope between the two points, and the trace will curve down if you decrease the velocity and curve upward when you increase it.

You will see this behavior in **Q11** through **Q14**. Increasing the velocity will create a concave up trace, while decreasing it will create a concave down trace. Switching from increasing to decreasing, but leaving the velocity the same sign, will cause your trace to have a point of inflection. Changing from a positive to a negative velocity will create a maximum in your trace maximum, while changing from a negative to a positive will create a minimum.

In a very real sense, what you are doing here is tracing the antiderivative of the velocity function you are creating by your movement of the velocity slider as time passes.

Explore More

Students can trace a function plot by adjusting the slope of the trace as they go.

Extension

To create a similar demonstration, construct a point in the plane (*Me2d*). Create a slider to represent time, and a slider to represent velocity. Add the time value to the x-coordinate of *Me2d*. Then, multiply the time value by the velocity value, and add this to the y-coordinate of *Me2d*. Plot these two new measurements as a point in the plane (*FutureMe*). Create a *Move* button to have *Me2d* chase *FutureMe*, and you're off and moving!

To create the particle itself to show motion along a line, create a line using the **Line** tool, and construct a point on this line (or use the x-axis and the origin, if you wish to model AP questions of this form). The coordinates of this point will serve as your origin. Add the y-coordinate of *Me2d* to the x-coordinate of your origin, and then plot this result and the y-coordinate of your origin as a point in the plane. A point will be created on your line whose position on the line is determined by the y-coordinate of *Me2d*, and will move as point *Me2d* moves.

Can You Predict the Trace? (page 11)

Prerequisites: Students should be able to describe the increasing and decreasing behavior of the graphs of various functions. The terms "concave up" and "concave down" are used informally.

Sketchpad Proficiency: Beginner/Intermediate. Students will need to measure coordinates and move objects in the plane.

Activity Time: 45–55 minutes

Document: Remotecontrol.gsp

General Notes: In this activity, students will experiment with simultaneous plots of the position and velocity of a point on a line and speculate about how the characteristics of each of the plots are related.

Sketch and Investigate

In **Q1**, since the velocity of point *Me* is left constant, the trace of the point *MyVelocity* will be a horizontal line while the trace of point *Me2d* will be a line, with a slope equal to the value of the velocity.

Moving the velocity slider after pressing the *Move* button will cause predictable changes in the position and velocity traces. In experiments A through E, you will find that the position trace will increase whenever the *velocity* slider is kept to a positive value. Thus, in **Q2**, the velocity trace will be positive whenever the position trace is increasing. In **Q3**, the velocity trace will be negative whenever the position trace is decreasing.

In **Q4**, changing the velocity value from positive to negative or vice versa will result in a maximum or minimum in your position trace, and your velocity trace will cross the x-axis.

In **Q5** and **Q6**, increasing the velocity value increases the "slope" of the position trace, and decreasing it decreases the slope. (Point *Me2d* is actually chasing a hidden point, S, along a line with a slope equal to the velocity value as time passes.) When the position trace is concave up, the velocity trace is increasing. When the position trace is concave down, the velocity trace is decreasing.

Using these characteristics, **Q7** asks for a prediction of a velocity trace given a position trace. Students may wish to observe the velocity values in order to assist in creating a paper and pencil sketch of the velocity plot, since the velocity trace will consist of time and velocity values. Eight examples to practice on follow, including one where the velocity is a physically impossible step function.

In trying to predict the position trace given a velocity trace, students can reconsider the connections between the traces. In **Q8**, when velocity is positive, the position trace will increase, although the trace need not itself be positive. This question is similar to **Q2**. In **Q9**, when velocity is negative, the position trace will decrease, although the trace need not itself be positive. This question is similar to **Q3**. Similarly, **Q10** through **Q12** mirror **Q4** through **Q6**.

Using these connections, **Q13** asks for a prediction of a position trace given a velocity trace. Students may wish to observe the actual motion of point *Me* in order to assist in creating a paper and pencil sketch of the position plot.

In the first example, it may be a surprise to students to see the velocity trace decrease while the position of point *Me* increases (point *Me* will move to the right). Seven additional examples to practice on follow, including one where the velocity is a physically impossible step function.

Explore More

Students are encouraged to change the location of point T, which sets the initial position of point *Me2d*. Since the velocity function does not change, the trace of position will be a vertical translation of the trace created with a different initial position.

Extension

The function that sets the velocity is hidden. On any of the sample pages, choose **Show All Hidden** from the Display menu. A function $g(x)$ will appear on the right side of the sketch window, along with other objects. Click on the function to deselect it, and then choose **Hide** from the Display menu to hide the other objects.

You can edit this function to change dependence of velocity on time, which allows you to model the type of question that begins "A point moves with velocity given by the function...." (You can even move point *Home* to the origin if you wish to model motion of a point along the x-axis.) Although the position trace in this activity is an approximation of the antiderivative of $g(x)$, no knowledge of this fact is needed to conduct the investigation, so the activity could be used at different times during a course for different purposes.

In this sketch, time will pass only when you press the *Move* button to make point *Me* move. It will "chase" point S, which is constructed so that a line through points *Me2d* and S has a slope equal to the value of $g(x)$ at the x-value of point *Me2d*.

Catching the Point (page 15)

Prerequisites: Students should be familiar with average velocity and slope as a measure of velocity.

Sketchpad Proficiency: Beginner/Intermediate. Students will need to measure coordinates, perform calculations in the calculator, and move objects in the plane.

Activity Time: 90–100 minutes. The activity will take 45–55 minutes if you start at step 11 on the page Easy Start.

Document: Motion.gsp

General Notes: Imagine watching an object in motion along a line. In observing its position, you might also note changes in its velocity. Suppose you were given control over the velocity of the object. Could you match the motion of the object with your velocity control? In this activity, students will try to match a plot of the position of an object by adjusting the velocity of another object.

Sketch and Investigate

In **Q1**, point Q will change direction at $time = 1.83$, 6.64, and 11.11. In **Q2**, point Q will appear to move with the least velocity at the times when it changes direction, and the greatest at the beginning and the end of the trip.

In **Q3**, point P, which plots the time and position of point Q in two dimensions, will move up whenever point Q moves to the right, and down whenever it moves to the left, because the y-coordinate of point P is equal to the position of point Q relative to point *Home*. In **Q4**, the vertical position of point P will change at the same rate at which the horizontal position of point Q changes.

The velocity of a point moving on a line, such as point Q, can be visualized by the slope of a line segment in the plane between its current time and position coordinates and its coordinates some time interval later. This line segment also represents a plot of the position of the point during that interval, assuming the velocity does not change.

In steps 6–8, students build segments in the plane with a given slope. Using this segment, they will try to match the plot of the position of point Q by adjusting the slope of the segment they create.

Point *FutureMe* is constructed by adding a small time value, $\Delta time$, to the x-coordinate of point *Me2d*, *mytime*, and by adding the change in position during that time to the y-coordinate, *myPosition*. The change in position is found by multiplying the change in time by the velocity.

In **Q5** and **Q6**, students experiment with this segment, observing that a smaller time interval produces a shorter segment and that changing the velocity value changes the slope of the segment. When students make point *Me2d* chase point *FutureMe*, a large change in the velocity will create a change in the slope of the trace left by the segment. In **Q7**, with a large time interval, the trace left will not accurately represent the position of a point moving along a line. Using a much smaller time interval, as in **Q8**, will produce a trace that can be used to try to match the trace created by point Q's motion along a line.

In **Q9** and **Q10**, using a positive velocity value causes your trace to go up, while a negative value causes it to go down. Switching from positive to negative causes a change in direction for the trace.

By plotting the velocity value (the slope of the segment between *Me2d* and *FutureMe*), students can see whether their attempts to match the position plot of point Q will also result in a match of the velocity of point Q.

Explore More

Students examine an impossible position plot. This plot requires an instantaneous change in velocity from -4 to 4 at $x = 4$ and from 4 to -4 at $x = 8$!

Extension

You may wish to construct a tangent line to the plot of point P's path to visualize the instantaneous velocity of point Q before your students attempt to match the plot with the "average velocity" segment between points *Me2d* and *FutureMe*.

To do so, select the position plot and choose **Properties** from the Edit menu. Select the Object tab and click on the top menu labeled Function g. Deselect the Hidden box. Then select **Derivative** g' from the Children menu. Deselect the Hidden box for this function as well. Click OK. Calculate $g'(time)$. Then choose **Plot New Function** from the Graph menu, and build the function $g'(time)(x - time) + g(time)$.

You may also wish to use the point *Me2d* to construct a point on the line containing point Q and point *Home* to see how your control of the velocity matches the location of point Q on the line as well. To do so, measure the distance from the origin to the unit point (1, 0). Multiply this value by the x-coordinate of point *Me2d*. Choose **Mark Distance** from the Transform menu. Then translate point *Home* by a vector of 0° and the marked distance.

Plotting Average Rates of Change (page 19)

Prerequisites: Students should be familiar with the idea of slope, step functions, and the average rate of change between two points of a function.

Sketchpad Proficiency: Beginner/Intermediate. Students will use custom tools, move objects in the plane, measure objects, and plot functions.

Activity Time: 50 minutes

Document: Step.gsp

General Notes: Calculations of the average rate of change of a function are often used as a way to introduce the concept and/or definition of the derivative. In this activity, students will plot the average rate of change between two points as a function and explore the properties of such plots as a precursor to investigations of graphs of derivatives.

Sketch and Investigate

In **Q1**, the average rate of change between points A and B is positive if the ratio Δy to Δx is positive (or when the segment AB slopes upward from left to right) and is negative if the ratio Δy to Δx is negative (or when the segment AB slopes downward from left to right). In **Q2**, point A can be on either side of point B. In **Q3**, the **Slope** tool will measure the slope of the line through point A and point B. The slope of the line through the points is the same as the average rate of change calculation because of the definition of slope.

After you merge the segment created in steps 1–5 to the function plot you created in step 6, you can measure the average rate of change of a function on adjacent intervals. In **Q4**, the average rate calculations will be equal because the rate of change of a linear function is a constant. Changes in the locations of the sample points or the lengths of the intervals will not change this result.

In **Q5**, the average rate of change of a quadratic function differs depending on the locations of the sample points or the length of the intervals. In **Q6**, on any interval where the function decreases, the average rate of change will be negative, and on any interval where the function increases, the average rate of change will be positive. If the function has equal values at two endpoints of an interval, then the average rate of change will be 0.

Because the slope of a secant segment is constant, the slope value can be plotted as a "step"—a horizontal segment—in the plane. In **Q7**, students can experiment with this step as a way to visualize the slope value as a function in its own right. The step is located below the x-axis where the slope of the line is negative, and above where it is positive. The step may move out of the window if the slope of the line is larger or smaller than the range of y-values shown on the vertical axis.

Just as the average rate of change of a function can be measured, the average rate can also be plotted as a function defined on a series of open intervals. In **Q8**, the steps appear to be a horizontal line because the average rate of change of a linear function is constant on any interval. In **Q9**, the steps do not move when you translate the function vertically, because the slopes are not changing. In **Q10**, the steps move up or down together to reflect the changes in the slope of the linear function plot, as they did with the segment in **Q7**.

In **Q11**, the plot of the steps for a quadratic function reflects the increasing (or decreasing, if the squared term is negative) behavior of the average rate of change. As in **Q9**, the steps do not move when you translate the function vertically in **Q12**.

In **Q13**, you can predict what the step function will look like given the function plot by estimating the slopes of a series of secant segments. Wherever the average rate of change of the function is positive, the step will be above the x-axis, wherever the average rate of change is negative, the step will be below the x-axis, and wherever the average rate of change is 0, the step will be on the x-axis.

Explore More

In **Q1**, students can quickly create a series of steps by tracing the location of one step whose length is fixed to investigate.

Note that the average rate is not defined at the x-values of the points you plotted, because average rate is not defined at a point. The question of rate of change at these points is a natural extension of this activity. You can use page 4 of the document to explore this question. As the value of h approaches 0, the step will decrease in length and will look like a plotted point, and the plot of the step function will look like the derivative plot.

Page 3 can also be used to motivate questions from students before the study of limits about whether one familiar function can be used to describe the rate of change of another. Try editing the function on page 3 to equal $\sin(x)$, for instance, making the value of h small, and pressing the *Move* button.

Going the Distance (page 22)

Prerequisites: Students should be familiar with the idea of velocity as a rate of change.

Sketchpad Proficiency: Beginner/Intermediate. Students will need to measure coordinates and move objects in the plane.

Activity Time: 35–45 minutes

Document: Area1.gsp

General Notes: In this activity, students will examine how area on a coordinate grid with units of time and velocity represents distance and how calculating the area can reveal changes in position.

Sketch and Investigate

In the document **Area1.gsp,** a function is plotted. In **Q1**, the velocity is constant for 6 minutes. In **Q2**, if you move at a constant velocity of 1 mile per minute for 6 minutes, you will travel 6 miles. In **Q3**, the area of quadrilateral *ABCD*, using the units of minutes and miles per minute, is also 6 miles. (Note that measuring the area of the quadrilateral in Sketchpad would not give a meaningful result in this context.) **Q4** introduces the idea that the area on a grid with units of time and velocity is a measure of distance.

Assuming constant acceleration, you can determine the average velocity over a time interval by finding the average of the initial and final velocities. Knowing the average velocity, you can determine the distance by multiplication. Finding the area of the triangle on the grid is equivalent. For **Q5** and **Q6**, you will again find that the area of the figure is the same as the distance—1 mile—found without any considerations of area.

Because the velocity does not change at a constant rate between 0 and 2 seconds, finding the area of a triangle (**Q7**) approximates the distance, but the result (1 mile (**Q8**)) is an overestimate (**Q9**).

In **Q10–Q15**, the total area of the triangle and the trapezoid is (1)(0.25) + (1/2)(0.25 + 1)(1) = 0.875 miles; using these two shapes gives a better approximation of the area between the function plot and the x-axis. The total area approximation using the previous answers is then 0.875 + 6 + 1 = 7.875. Using the area tools, the actual area from 0 to 2 is 2/3 mile. The actual area from 0 to 10 is 7 2/3 miles.

Explorations 1 and 2

The area calculation provided in the sketch depends upon a hidden calculation of the integral of the function *f*. The activity "Slope and Limits," in the Exploring Limits chapter, deals with how to guess the area of a region involving a curve in more detail. Students can use additional trapezoids to see why the sketch gives the value of 0.66667 as the area under the curve from 0 to 2.

Using the four subdivisions, the area approximation will be 0.6875. %*Error* = 0.03125.

For this function, the least amount of error using the four subdivisions is found when four equal subdivisions are used.

Extension

The measurement of the area between the function plot and the x-axis, as mentioned above, is taken by substituting the x-value of point *P* into another, hidden, function. This function evaluates the integral of *f* from point *start* to point *P*. Point *P*, as you have seen, can move, so you get a dynamic area calculation as you move point *P*.

You can build a similar demonstration, where you display the value of the area as the area "fills in" in at least two ways.

One way is to create a new function, calculate its derivative, and plot this derivative function. Then construct a point *P* on the x-axis, and measure its x-coordinate. Find the value of $f'(x_P)$, and plot $(x_P, f'(x_P))$. Construct a segment between these two points. By animating or moving point *P* and tracing the segment, you can show the area between your plot and the x-axis being filled in, starting from a particular location.

To show the area being accumulated, you first need a starting point. Construct another point on the x-axis, label it *start*, and measure its x-coordinate. Then, create a new function equal to $f(x) - f(x_{start})$. The value of this function for any x-value is equal to the accumulated area between the plot of $f'(x)$ and the x-axis, starting at point *start*. By showing the value of the function *f* as you do this, you'll be able to show how the area accumulates and what the total area is.

In the document provided with this activity, the same result was accomplished by first plotting a function, then determining its antiderivative, and displaying the value of the antiderivative at $x = x_P$ minus the value of the antiderivative at $x = x_{start}$.

What Do You Expect? (page 26)

Prerequisites: Students should be able to work with functions graphically, numerically, verbally, and symbolically.

Sketchpad Proficiency: Beginner/Intermediate. Students will need to measure coordinates and move objects in the plane.

Activity Time: 45–55 minutes

Document: Limits.gsp

General Notes: In this activity, students will investigate the concept of limits from graphical and numerical perspectives by examining function plots, and dynamically adjusting function values and scales.

Sketch and Investigate

In **Q1** and **Q2**, students may be able to move the points *Friend* and *Me* so that their x-values are less than 0.01 from $x = 2$, suggesting the value of the function at $x = 2$ is 12. In **Q3** and **Q4**, the limit of the function as x approaches 2 is 12 from both sides.

In **Q5** through **Q7**, students' answers will vary, depending on how far they zoom in and how close points *Friend* and *Me* were to $x = 2$ to begin with. Zooming in, however, supports the idea that the function has a limit of 12 as x approaches 2.

In **Q8** and **Q9**, students might expect that the limit as x approaches 0 is 1, and this should be supported by zooming in.

In **Q10** through **Q14**, students might expect that the limit as x approaches 2 is 6. Upon zooming in, however, this expectation may change. The values of the function as x approaches 2 approach different values on the left and right sides. From the left, the function approaches a value of 5.99, while on the right, the function approaches 6.01. The function cannot take on values between these two numbers, so 5.995 cannot be achieved. Because the left- and right-hand limits are not equal, the function does not have a limit as x approaches 2.

Explore More

Students can examine a limit as x approaches infinity by using the sliders to zoom out. In **Q1** through **Q4**, the value of the function at $x = 30$ is approximately -0.03, and as x is made larger, the function will continue to have values closer to 0. By $x = 300$, the value of $f(x)$ is approximately -0.003.

You can also use page 4 of the document to investigate any limit you wish. Just change $q(x)$ to the function that you'd like to examine, change a to the x-value you are approaching, and change b to your initial guess for the limit.

Extension

If you do not wish a point to appear at the location where points *Friend* and *Me* seem to converge, do not delete the point labeled "?" but hide it and the objects that depend upon it. (Select the objects you wish to hide, and then choose **Hide Objects** from the Display menu.) Note, however, that this point is the point around which the axes will zoom. To change the location of this point, edit the values of a and b by showing the **Zoom** tools and editing the values of a and b with the calculator.

The **Zoom** tools in this activity can be used to zoom in on any point in the plane. Change the values of a and b to zoom in on a new point. Note that the value of a must be determined by a value you type in; it cannot be a value of some other object in the sketch, because much of the sketch is constructed using this value.

You can, however, edit the value of b to equal $q(a)$ if you wish, as long as the function q is defined there. Otherwise all your axes will disappear, because the value of the point upon which the axes zoom is determined by the values of a and b.

Although the examples included in the activity focus on places where the function is undefined, you can add to this activity by including a function where there is no discontinuity in the function and zoom in on any point. Edit the function and edit the values of a and b to zoom in on a point of your choosing.

Although there is another activity in the chapter "Exploring Limits" concerned with exploring the epsilon and delta definition of limits, you can use the vertical and horizontal segments included here to generate questions and discussion about how close you must make points *Friend* and *Me* to a particular x-value to ensure that the function lies within a particular range on the y-axis. The function on page 3 of the document is particularly suited to this, as the "gap" in the function is not visible unless you zoom in.

In the sketch, points *Friend* and *Me* are free to move along the function plot. You may wish to control the x-distance from these points to the x-value at which you wish to investigate the limit. To do so, construct a slider and label the value of the slider h. Use the calculator to calculate $c + h$ and $c - h$. Then calculate $q(c + h)$ and $q(c - h)$. Plot the points $(c + h, q(c + h))$ and $(c - h, q(c - h))$.

Continuous or Discontinuous? (page 33)

Prerequisites: Students should be able to work with functions graphically, numerically, verbally, and symbolically.

Sketchpad Proficiency: Beginner/Intermediate. Students will need to plot points, measure coordinates, and move objects in the plane.

Activity Time: 45 minutes

Document: Continuity.gsp

General Notes: In this activity, students will investigate continuity using interactive function plots.

Sketch and Investigate

Hiking along the function plot wouldn't seem to present any surprises.

In **Q1**, a discontinuity occurs at $x = 2$. In **Q2** and **Q3**, as you drag point A to the right, point Q is on the function plot until point A reaches $x = 2$, where point Q disappears. The y-coordinate of point Q is at most 4. In **Q4**, you should be able to get as close as 1.99999 and 3.99999, or even 2.00000 and 4.00000. These values are still before the continuity, however, because if you were to move point A yourself, reaching $x = 2$ would cause point Q to disappear. In **Q5**, the limit of the y-coordinate of point Q is 4.

In **Q6** through **Q9**, you should find the coordinates (2.01, 3.455), (2.01, 3.27005), and (2.001, 3.25201). The limit on this branch is 3.25 (**Q11**).

In **Q12** through **Q14**, the function approaches positive infinity from the left and negative infinity from the right. From the left, you should find the coordinates (–1.1, 1.22), (–1.01, 10.22), and (–1.001, 102.2.) From the right, you should find the coordinates (–0.9, –0.82), (–0.99, –9.8), and (–0.999, 99.8).

In **Q15** through **Q18**, the three types of discontinuities are called (in the order they appear): a point discontinuity, a step or jump discontinuity, and an asymptotic or infinite discontinuity. A close inspection of the discontinuity that felled the hiker would reveal it to be two step discontinuities.

Explore More

1. all integer values, jump
2. $x = 1$, infinite
3. $x = 2$, point
4. $x = 0$, point
5. $x = 0$, infinite
6. $x = \pi/2 + k\pi$, where k is any integer, infinite
7. $x = 2$, jump

Extension

There are a number of ways to plot discontinuous functions in Sketchpad. You may be familiar with some of the methods. Adding the function

$$g(x) = 0.5\left(\frac{|x-a|}{x-a} + 1\right)$$ to a function f, for instance,

will add 0 where $x < a$ and a where $x > a$, creating a step discontinuity, and the function will be undefined at $x = a$.

After plotting $f(x) + g(x)$, remember to go to **Edit | Properties | Plot** and choose **Display as: Discrete.**

A function with a step discontinuity that *is* defined at $x = a$ can be created by adding the function $g(x) = \text{sgn}(1 + \text{sgn}(x - a))$ to $f(x)$. This function will add 0 where $x < a$, and 1 where $x \geq a$. To create a function with a discontinuity at a point, you can use familiar functions such as $f(x) = (x^2 - 9)/(x - 3)$.

The case where a function is discontinuous at a point, but is defined at that point, illustrates that the value of a function at $x = a$ does not necessarily affect the value of the limit as x approaches a.

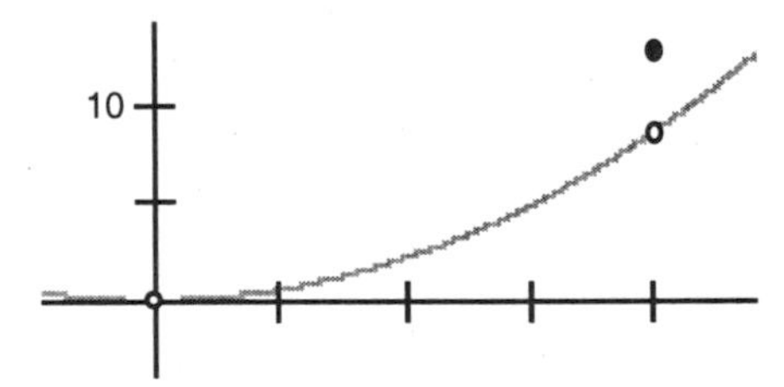

You can create such a function in Sketchpad by adding the function $g(x) = \text{sgn}(x-a) \cdot \text{sgn}(x-a) + 1$ to a function f that is defined at $x = a$. Adding this function will add 1 to $f(x)$ at $x = a$ only, creating the situation shown in the diagram above. (You can also add expressions like $3g(x)$ or $-2g(x)$ to have your defined point appear at a greater or lesser y-value.)

When you plot $f(x) + g(x)$, the defined point will not appear. If you evaluate the function at $x = a$, however, you can then plot the point $(a, f(a) + g(a))$ to create the display shown above. (Another point was plotted, and then its color was set to white to give the appearance of a "hole." This could be done for any function where you want to illustrate that type of discontinuity.

See the document **Discontinuities.gsp** for some completed examples.

Delta, Epsilon, and Limits (page 38)

Prerequisites: Students should have an informal notion of the concept of limits.

Sketchpad Proficiency: Beginner. Students will move objects in the plane, and they need to be comfortable with resizing windows.

Activity Time: 30–45 minutes

Document: EpsilonDelta.gsp

General Notes: In this activity, students will explore the meaning of the epsilon-delta definition of the limit of a function from a graphical and numerical perspective. The document includes a "green light" indicator that will signal students when they have selected a value of δ such that when x is within δ units of a number c ($x \neq c$), the y-value of a function f is within a given distance (ε units) of a number L.

Sketch and Investigate

In **Q1**, a value of $\delta < 0.04$ is sufficient to ensure that the y-values of the function are within ε of the value of L when x is within δ units of c. (In Case 1, the value of c is 2 and the value of L is $f(c)$.) Any smaller value of δ will also work.

In **Q2**, you will find that if ε is set to 1/2 of its previous value, δ will also need to be halved so that the "green light" indicator signals that the value of the function is within ε of the value of L when x is within δ units of c. The same relationship will hold for 2/3 or 2/5.

For **Q3**, in each of the cases on the first page of the document, it is possible to set a value of δ that satisfies the definition of a limit at $x = c$. For **Q4**, you may find that you cannot use the slider for δ to ensure that the function is within ε of the value of the limit. On page 1, this limitation is the result of the values that the slider takes on, and does not indicate that the limit does not exist. The fact that this occurs can motivate discussion about the need for a proof that the value of δ exists for any ε, no matter how small, in each case.

In **Q5**, on page 2 of the document, the limit of $f(x)$ will fail to exist in the case where x approaches 1, as the function is not defined there.

Extension

You can use the functions shown to have students predict algebraically when the indicator light will change. Have students construct an inequality using the function and the two values between which you want the function to lie, and solve the inequality. By using ε instead of a particular numerical value, you can construct a proof as mentioned above.

Example: If $f(x) = x^2 - x + 2$, how close to 2 must x be in order to insure that $f(x)$ is within 0.1 of 4?

Inequality: $3.9 < x^2 - x + 2 < 4.1$

By solving this inequality, you can find the needed value of δ. More generally, solving $3 - \varepsilon < x^2 - x + 2 < 3 + \varepsilon$ will tell you the relationship between the value of ε and the value of δ. You can use the document **EpsilonDelta.gsp** to visualize the solution to the inequality.

To visualize an epsilon-delta proof, you could also construct the value of δ so that it depends in some way upon ε. To do this, delete the slider for δ. The rectangles will also be deleted, because they depend on the measurement associated with the slider, as will the indicator circle. Measure the x- and y-coordinates of point C.

Suppose you have proved that for a certain function, δ must always be less than $\varepsilon/5$. Use the calculator to calculate $\varepsilon/5$, and add and subtract this calculation to the x-coordinate to create two new measurements. Use these measurements to plot points on the x-axis to the left and right of point C. Evaluate the function at these x-values to create points on the function plot (using the **Plot as (x, y)** command). Construct lines perpendicular to the y-axis through the points on the function plot, and construct the intersections of these lines with the y-axis. You can then reconstruct the rectangles showing the region in which the function's values lie.

Given any ε, the value of δ will always ensure that the y-value of the function f is within ε units of the number L. As you change the value of ε with the slider, the value of δ will remain equal to $\varepsilon/5$, and the rectangle will stay within ε of the limit.

For a different function at a given point, the value of δ may differ. To create a different demonstration, edit the expression for $\varepsilon/5$ to the value that results from your proof, and the rectangles will be changed accordingly. You can then again demonstrate that your choice of δ will ensure that the function's values lie within ε of the limit.

How Close Do You Go? (page 40)

Prerequisites: Students should have an informal notion of the concept of limits.

Sketchpad Proficiency: Beginner/Intermediate. Students will move objects in the plane.

Activity Time: 40–50 minutes

Document: Limits2.gsp

General Notes: In this activity, students will use the delta-epsilon definition of the limit to examine limits where the function is discontinuous, as well as other situations involving limits.

Sketch and Investigate

In **Q1**, the function is undefined at $x = 3$ because that value gives the indeterminant form $0/0$. In **Q2**, the plot suggests that the function has a limit of 2 as x approaches 3. The function f is equivalent to $(x + 3)/3$ for $x \neq 3$, which has a value of 2 at $x = 3$. In **Q3**, to make the y-values of the function fall within the interval from 1.5 to 2.5, you must be within 1.5 of c. In **Q4**, to make the y-values fall within 0.1 of 2, you must be within 0.3 of c. In **Q5**, to be within 0.05 of 2, you must be within 0.15 of c, and in **Q6**, within 0.03 and 0.003. For any ε, you must be within $\delta = 3\varepsilon$ units of c on either side; the y-values are symmetric around $x = 3$ (**Q7** and **Q8**). In **Q9**, x must be within 0.13 of $c = 1.5$ on the right side so that the y-values are within 0.1 of 1.5. On the left side, x must be within 0.16 of $c = 1.5$. To make the y-values fall within the desired range, the smaller value (0.13) must be used.

In **Q10**, the function has a point discontinuity at $x = 3$. Although the limit of the function is 9, the value of the function at $x = 3$ is 10. For **Q11**, x must be within 0.016 of $c = 3$ on both the left- and right-hand sides; the intervals will appear to be symmetric, (because the function plot closely approximates a line when you zoom in) although they are not, as the function has a squared term.

In **Q12**, the function is not continuous at $x = 3$ because of the step discontinuity there. Students may initially guess that 9 is the limit as x approaches 3, but in **Q13** and **Q14**, though the function is defined at $x = 3$ ($f(x) = 10$) the value of the function approaches 9 from the left and 10 from the right. There is no value of δ that will ensure that the function lies within, say, 0.1 of 9 on the left, because the "gap" in the function is larger than this interval. To visualize this, students will find that the horizontal band cannot be made any smaller than a width of 1. Because you cannot make the y-value of the function as close as you wish to one particular number L by making x sufficiently close to c, the limit of the function does not exist at $x = 3$.

Explore More

The example in step 1 presents a situation similar to the previous examples in all respects except that the function is not defined at $x = 3$. The limit as x approaches 3 does not exist for the same reason as the previous example.

The examples in steps 2 and 3 present different dilemmas. For $f(x) = 1 + x\cos(1/x)$, $f(x)$ oscillates with decreasing amplitude as x approaches 0. The function $f(x) = 1 + \cos(1/x)$, however, oscillates with an amplitude of 1 as x approaches 0.

In the first case, the y-value of the function can be shown to lie within any given small interval around 1 by making x sufficiently close to 0. In the second case, however, no matter how close x is to 0, the function will still oscillate between 0 and 2, thus preventing you from making the y-value of the function as close as you wish to any one particular number L.

As x becomes infinitely large, $f(x) = 1 + x\cos(1/x)$ will also grow infinitely large because the value of x continues to grow while $\cos(1/x)$ approaches 1. As x becomes infinitely large, $f(x) = 1 + \cos(1/x)$, on the other hand, approaches a limit of 2, again because $\cos(1/x)$ approaches 1.

Extension

When a function has a limit as x approaches infinity, the difference between the value of the function and the limit L can be made as small as you wish by making x sufficiently large.

To visualize this concept with Sketchpad, construct the plot of a function you suspect might have a limit as x grows infinitely large. Then construct a point on the x-axis, measure its coordinate, find the value of $f(x)$ for your point, and plot the point $(x, f(x))$. Construct a line through this point, parallel to the x-axis. Move this point out on the x-axis, and use the line to observe the value of the function as x grows larger. How large do you need to make x to make the value of the function within 0.1 of the limit? Within 0.01?

If you wish to investigate such a function, you can use the **Zoom** tools to zoom around the point (a, b). By dragging the x slider to the far left (move the slider itself to the right so that you can increase the scale) your x-axis can range into the millions, and you can observe the value of the function as your point moves toward these large values.

Slope and Limits (page 44)

Prerequisites: Students should be able to work with functions graphically, numerically, verbally, and symbolically and should be familiar with the concept of limits.

Sketchpad Proficiency: Beginner/Intermediate.

Activity Time: 50 minutes

Document: SlopeandLimit.gsp

General Notes: In this activity, students will examine how limits can be used to ask what at first might seem like an impossible question: What is the "slope" of a function that is not linear? In this activity, students begin by examining the rate of change between two points on the plot of a linear function. When they edit this function so that it is no longer linear, they will find that the concept of limits allows for an easy leap from the slope of a line to the slope of a curve.

Sketch and Investigate

In **Q1**, moving point Q will change the value of *slopePQ*; as you move point Q to the right, the value of *slopePQ* will increase. Because point Q is free to move along the function plot, it has coordinates (x, x^2). In **Q2**, *slopePQ* is then $(x^2 - 9)/(x - 3)$. The limit of this expression (which is a function in its own right) as x approaches 3 is 6 (**Q3**). The numerical calculation of the slope in the sketch should support this value—as point Q approaches point P, *slopePQ* will approach a limiting value of 6 (**Q4** and **Q5**).

The limit of function can be visualized by examining the y-value of the function near a particular value of x. In examining the limits above, however, it is the slope of a line that provides the visualization of the limit. In **Q6**, examining the slope on a zoomed-in portion of the function plot results in a value that does not change. Although the function $f(x) = x^2$ is not linear, a sufficient amount of zooming creates the appearance of linearity. One answer to **Q7** recognizes the fact that the slope of the line PQ changes less and less as point Q approaches point P.

Exploration 1

In **Q1**, the length of the segment is a way to visualize the error between the actual function value and the value of the secant line at a particular value of x.

In **Q2**, the error at $x_R \approx 2$ is approximately 1.

In **Q3**, to reduce the error to 0.1, x_R should be approximately 0.316 from 3. To reduce the error to 0.05, x_R should be approximately 0.224 from 3. To reduce the error to 0.001, x_R should be approximately 0.316 from 3. In **Q4**, the equation of this line would be $L(x) = 6(x - 3) + 9$. In **Q5**, $L(3.04) = 9.24$ and $f(3.04) = 9.2416$, so the error at $x = 3.04$ is 0.0016.

Exploration 2

In **Q1** and **Q2**, the left-hand limit of the secant line is -6, while on the right, the limit is 6. The $\lim\limits_{x \to 0} \dfrac{f(x) - f(0)}{x - 0}$ does not exist. Although you do not have a plot of the function $\dfrac{f(x)}{x}$, you can tell from the behavior of the lines that the limit of this function as x approaches 0 does not exist. This behavior means that the function $f(x)$ cannot be approximated by one line near $x = 0$, and so the function is not locally linear there (**Q3**).

Exploration 3

Although the functions in this exploration do not have "slope," the use of the secant line as an approximation for the function, along with the concept of limits, allows you to say the function has a slope at a point—the instantaneous rate of change.

In **Q1**, the limit as x approaches π of the expression $(\sin(x_P) - \sin(\pi))/(x_P - \pi)$ would determine the instantaneous rate of change of the function $\sin(x)$ at $x = \pi$. In **Q2**, the limit as x approaches 4 of the expression $(\ln(x_P) - \ln(4))/(x_P - 4)$ would determine the instantaneous rate of change of the function $\ln(x)$ at $x = 4$. In **Q3**, the limit as x approaches 4 of the expression $\left(\sqrt{x_P} - \sqrt{4}\right)/\left(x_P - 4\right)$ would determine the instantaneous rate of change of the function $\ln(x)$ at $x = 4$.

In **Q4**, the instantaneous rate of change of a function will not exist whenever the slope of the line from a given point close by does not approach a particular value. This may happen where the graph of the function has an abrupt change in direction or where the graph itself is discontinuous. The absolute value function is an example where an abrupt change in slope occurs. The slope of the line used to visualize the instantaneous rate of change may also fail to have a limit where the graph becomes vertical—the graph of $f(x) = \sqrt{x}$ at $x = 0$, for instance.

Area and Limits (page 49)

Prerequisites: Familiarity with the concept of limits.

Sketchpad Proficiency: Beginner/Intermediate. Students will move objects in the plane.

Activity Time: 50 minutes

Document: AreaLimits.gsp

General Notes: Students may be familiar with the idea of approximating the area of an irregularly shaded region by "counting boxes." In this activity, they will extend this idea by using rectangles. By adjusting the width of these rectangles in a dynamic construction, students will explore how limits can be used to find the area of a region.

Sketch and Investigate

In **Q1**, the shaded region is a rectangle, so the area is $10 \cdot 4.9$, or 49 square units. In **Q2**, the area is $0.5(4.9 + 8.9)(10)$, or 69 square units. In **Q3**, the area of the trapezoid on page 2 is larger than the area of the shaded region on page 3. In **Q4**, students will likely make an estimate near 47 square units using whole boxes, and it is an underestimate. Students may have varied methods for how to count the partially shaded boxes. For **Q5**, students may suggest that a finer grid, using boxes of 0.5 units, for instance, would allow a more accurate measurement. They can do this by moving the unit point to the right until there are grid lines every 0.5 units. Students may want to get a better estimate by moving the point further to the right until there are grid lines every 0.25 units. They can also devise a method for counting the partial squares.

In **Q6**, students can use a calculator to add the y-coordinates of point C and get the value of 50.5, which is an underestimate. In **Q7**, the sum of the right rectangles can be found using a calculator with a table feature to quickly generate the values of f at $x = 1, 2, 3$, and so on. Finding the sum of the values for $x = 1$ to $x = 10$ results in the value of 54.5. This is an overestimate.

In **Q8** through **Q11**, if $n = 1$ the difference between the estimates is 40, but if $n = 10$ it is only 4. If $n = 20$, the difference is 2. For $n = 25$, it is 1. In general, the difference between the two sums is $40/n$.

In **Q12**, you can use the "difference between estimates" rectangle to visualize how the difference is related to n. The left and right sums share all but the leftmost and the rightmost rectangle for any value of n, so when the sums are subtracted, all that is left is the difference between the areas of these two rectangles. In **Q13**, for any value of n, this difference is $f(b)\left(\frac{b-a}{n}\right) - f(a)\left(\frac{b-a}{n}\right) = (f(b) - f(a))\left(\frac{b-a}{n}\right)$. Because $(f(b) - f(a))$ and $b - a$ are fixed values, as n is made larger, the difference can be made as small as you wish. As n approaches infinity, the two estimates converge.

Explore More

In **Q1**, the exact area of the region is 55 square units. For **Q2**, the area of the 5 left sum rectangles is $\frac{10}{5}\sum_{i=0}^{4}\left(0 + \frac{10}{5}i\right) + 1$. In **Q3** though **Q5**, for any value of n, the width of the rectangles is $10/n$, so the left sum area can be calculated using the expression $\frac{10}{n}\sum_{i=0}^{n-1}\left(0 + \frac{10}{n}i\right) + 1$. The exact area of 55 is equal to $\lim_{n\to\infty}\frac{10}{n}\sum_{i=0}^{n-1}\left(0 + \frac{10}{n}i\right) + 1$.

For $f(x) = 0.1(x - 3)^2 + 4$, the expression for 20 rectangles is $\frac{10}{20}\sum_{i=0}^{19}0.1\left(0 + \frac{10}{20}i - 3\right)^2 + 4$. One way to find the value of this expression is to expand it and then apply the formulas for the sums of sequences.

Extension

To show a different example using this document, you can adjust the endpoints of the intervals on which the area is shown on any page by editing the parameters a or b. The expression for f can of course also be edited.

Extra for Sketchpad Experts: To create the shaded regions, a locus was used, so the area of the shaded region cannot be measured directly, even when it is a rectangle or a trapezoid. The rectangles for the left and right sums are an iterated image of the rectangle on the far left of the interval.

The area calculation was created by constructing a point and then adding the value of this first rectangle to its y-value. Iterating this point, and measuring the y-value of the terminal point of the iteration, gives the sum of the area of the rectangles. This construction will be used in the later activities concerning integrals.

Taking It Near the Limit (page 55)

Prerequisites: Students should be (informally) familiar with the concepts of limits and average rates as slopes of secant lines.

Sketchpad Proficiency: Beginner/Intermediate. Students will need to measure coordinates, use the calculator, and move objects in the plane.

Activity Time: 45–55 minutes

Document: Derivative1.gsp

General Notes: In this activity, students will investigate a definition of the derivative by zooming in on a secant line at a given point on a function plot.

Sketch and Investigate

In **Q1**, answers may vary, as students will construct their own points. In step 7, if you wish to have one of your points at the point (2, –2), plot this point and then create a *Move* button to move point P to this location.

In **Q2** through **Q3**, moving point Q close to point P to approximate the limit of the slope of the secant line should result in a value close to 1. If you were to move point Q to the same location as point P (which you can do with a *Move* button), the average rate calculation would be undefined.

When you zoom in, the coordinates of points P and Q will not change. If you have not moved point P to the location (2, –2), you can move it closer once you have zoomed in.

In **Q4** through **Q6**, moving point Q closer to point P should further support the idea that the limit of the secant slopes is 1. So that the points P and Q do not appear to separate themselves from the function plot at a highly zoomed-in scale, you can change the domain of the function plot before you zoom in to a small domain around $x = 2$.

In **Q7**, if students are familiar with the term *tangent line,* they may use it here. More generally, upon zooming out, the secant line will appear to intersect the function plot only once, and its slope is a visual representation of the (almost) instantaneous rate of change of the function at $x = 2$.

In looking for a local maximum, you may wish to zoom in gradually and relocate point P after making an initial guess for the x-value of the maximum. You can edit the values of a and b to be equal to your guess to make zooming in easier.

In **Q8** through **Q11**, a value of x a short distance to the right of $x = 0.18$ will yield a maximum for the function. As you zoom in, the function will appear flatter, and the slope of the secant line (and the limit of the average rate of change) will be 0.

Explore More

The only change for **Q1** is that point P has a fixed value (in this case, π), and that does not affect the limit or average rate. As mentioned above, you may wish to plot a fixed point on the function plot rather than construct a moveable one in order to make zooming in easier.

In **Q2**, for 3a and b, the slope of the secant line will change abruptly at the points indicated. In 3c, zooming in will lead to an ever increasing slope value for the secant line. In each of these cases, the derivative does not exist because the limit of the slope of the secant line does not exist.

Extension

You can use this sketch to investigate the derivatives of a variety of functions at a point by simply editing the expression for f. To investigate the derivative at a particular point on a function plot, edit the expression for b so that it equals $f(a)$. To do this, double-click on b, delete the value, and enter $f(a)$ by clicking on the function f, then on the value a, and then closing the parentheses. You can change the value of a at any time.

Plot the point (a, b) on the coordinate grid by selecting a and b and choosing **Plot as (x, y)** from the Graph menu. This point will now be a point on the function plot. Use a line between this plotted point and a point Q constructed on the function plot as in the activity to investigate the derivative.

To demonstrate another definition of the derivative, build a slider in the following manner. Create an independent point in the plane, S, and translate it by the vector from the origin to the point (1, 0). Construct a ray from point S through point S', and construct a point, T, on this ray. Select S, S', and T, and choose **Ratio** from the Measure menu. Label this ratio h, and calculate $a + h$, $a - h$, $f(a + h)$, and $f(a - h)$. Plot the points $(a + h, f(a + h))$ and $(a - h, f(a - h))$. Use point P and the two new points to investigate the symmetric difference quotient and examine its limit as $h \to 0$ by adjusting your slider. Or examine the limit of $(f(a + h) - f(a))/h$ as $h \to 0$ by examining the slope of the secant line drawn to $(a, f(a))$ from either direction as you adjust the slider. You can zoom in as before and continue to use your slider for h. (Feel free to relabel a or h to conform with your preferred notation.)

Going Off on a Tangent (page 59)

Prerequisites: Students should be familiar with representing average rates as secant slopes, and with the idea of the derivative as the limit of the average rate of change.

Sketchpad Proficiency: Beginner/Intermediate. Students will need to measure coordinates and move objects in the plane.

Activity Time: 45–55 minutes

Document: Tangents.gsp

General Notes: In this activity, students will visualize the instantaneous rate of change of a function as a tangent line to the plot of the function and use the slope of the tangent line to speculate about the derivative.

Sketch and Investigate

In the document **Tangents.gsp**, point P is a point on a function plot. A tangent line has been constructed to the plot of f through point P by calculating the derivative of f (using the **Derivative** command) at the x-value of point P and using the derivative value to plot the equation of the tangent line through point P. These calculations are hidden, as the activity assumes only that students are familiar with the derivative as a limit of average rates of change.

In **Q1**, **Q2**, and **Q3**, students can move point P to visualize and estimate the instantaneous rate of change of the function (the derivative) by using the grid to estimate the slope of the tangent line. At $x = -1$, the slope of the tangent line appears to be 10; at $x = 0$, 1, and at $x = 1$, –2.

In **Q4** and **Q5**, relationships between the increasing behavior of the function and a positive slope of the tangent line and decreasing behavior of the function and a negative slope of the tangent line are visualized. When the derivative is positive, the function is increasing, and when the derivative is negative, the function is decreasing. **Q6** introduces the idea that when a function has a maximum or a minimum the derivative equals 0. (You may wish to edit the function f to illustrate that a function's derivative can equal 0 without a maximum or minimum occurring.) In **Q7**, the derivative is greatest at both endpoints of the interval $[-1, 3]$.

Q8 repeats Q4–Q7 for a sine function, while asking students to use what they have seen to visualize the tangent line at different locations on the function plot. By examining where the function is increasing, decreasing, or 0, students can estimate the value of the derivative.

In **Q9–Q11**, students can explore the connection between the slope of the tangent line and the curvature of the function plot. Whenever the tangent line is "under" the function plot, the slope of the line is increasing, and the function plot is concave up.

Explore More

The three functions listed have a point or points where the derivative is undefined. For the absolute value function, the tangent line can have only a slope of –1 or 1. Both this function and the next exhibit an abrupt change in slope at particular x-values. With the second and third functions, however, the abrupt change is not of the same nature. For both these functions, the slope of the tangent line increases without bound. This can be seen by zooming in and continuing to observe the slope.

Extension

You can plot points on the function plot and create buttons that will move point P to specific locations. To do this, use the calculator to calculate a value of 1. (All you need to do in the calculator is type in the number 1.) Then calculate $f(1)$ and plot $(1, f(1))$ as a point in the plane. Finally, create a button to move point P to the newly plotted point.

To examine the local behavior of the function near any particular point, you can use the zoom controls included in this sketch.

To make this process easier, plot (a, b) as a point in the coordinate grid. Create a button to move point P to this point. After pressing the *Move* button you have created, point P will be at the exact location around which the sketch will zoom. Then use the **Zoom** tools to examine the behavior of the function near the point. As you zoom in, the function plot should become indistinguishable from the tangent line—except in the case of the functions listed in the Explore More section.

Note for advanced Sketchpad users: You can't edit the values of a and b to equal other measurements in the sketch because much of the sketch is dependent on the values of a and b.

Plotting the Derivative (page 62)

Prerequisites: Familiarity with the concept of the derivative of a function at a point.

Sketchpad Proficiency: Intermediate. Construction of measurements, plotting points, functions, and moving objects.

Activity Time: 50 minutes

Document: PlotDerivative.gsp

General Notes: In this activity, students will plot approximate derivative values using the slope of a dynamic secant line. By moving this line, they will explore how to create a derivative plot.

Sketch and Investigate

In **Q1**, the secant line that you construct is an approximation of the derivative at $x = x_P$. (Note that this is an arbitrary choice. The line could also be used to approximate the derivative at $x = x_Q$.) Finding the value of the derivative at $x = x_P$ requires finding the limit of the slope of the line PQ as point Q approaches point P. By moving point Q as close as possible to point P, you can minimize the error in the approximation of the derivative.

By placing the points close together and moving them along the function plot, the locations of point S trace the approximate value of the derivative at $x = x_P$. Note that the difference between the x-coordinates of points P and Q may change as you move them, so the precision of the approximation may change slightly as you move the points.

In **Q2**, the trace created is a parabola.

The trace will differ according to whether $x_Q > x_P$ or $x_Q < x_P$. This difference, as suggested in **Q3** through **Q5**, is a result of the slope of the function not being symmetric around point P. The discrepancy is greater if the points are further apart, and less if they are closer. If $x_Q > x_P$, you are calculating the limit in the derivative from the right side, sometimes called the *right-hand derivative*. If $x_Q < x_P$, you are calculating the limit from the left.

In **Q6**, students can adjust h to make the locus match the trace they made with the *Animate* button. The traces will match when the value of h is equal to the difference between the x-coordinates of points P and Q when the *Animate* button was used.

Symbolically, $h = x_Q - x_P$. The value of h can be positive or negative, so you can match either the trace created when $x_Q > x_P$, or when $x_Q < x_P$.

Explore More

Students are asked to predict what the plot of the derivative will look like by examining the slope of the tangent line. For **Q1**, point S will have a negative y-coordinate if the slope of the line is negative, a positive y-coordinate if the slope of the line is positive, and a y-coordinate of 0 if the line is horizontal.

In **Q2**, focus first on the actual slope of the line and then the changes in the slope of the line. If students have difficulty with this question (they may make the error of thinking that the slope of the line is decreasing when it is actually the value of the function that is decreasing), you can ask them to think about how to determine when point S will be moving up and when it will be moving down. For instance, when the slope of the line decreases, the point will move down. When the slope of the line increases, the point will move up. In **Q3**, the locus is a cubic.

Extension

This document provides an opportunity for students to move point P themselves and draw conclusions about the motion of point S (and the increasing/decreasing behavior of the derivative) before trying to sketch a derivative graph on their own. Also consider having students record their observations and conclusions in writing, to help them make precise verbal descriptions of the relationships between the plot of a function and the plot of its derivative.

As another activity, have students create a independent point in the plane, and turn on tracing for this point. With point S hidden, run an animation of point P, have students move this independent point to try to trace the plot of the derivative. Then, without erasing the traces, run the animation again with point S showing and its trace on, and have students compare their trace with the trace created.

To explore a symmetric difference quotient follow these steps:

1. Calculate $x_P - h$ and $f(x_P - h)$.
2. Plot the point $(x_P - h, f(x_P - h))$. Label this R.
3. Construct the line QR and measure its slope (instead of PQ).
4. Select x_P and the slope measurement, *slopeRQ*, in that order. Choose **Plot As (x, y)** from the Graph menu.

So What's the Function? (page 66)

Prerequisites: Familiarity with the derivative of a function at a point, as well as the effects of combining functions with elementary operations.

Sketchpad Proficiency: Beginner/Intermediate

Activity Time: 50 minutes

Document: DerFunction.gsp

General Notes: In this activity, students will plot an approximation of the derivative by plotting the function $(f(x + h) - f(x))/h$ and make conjectures about derivatives and their properties using this approximation.

Sketch and Investigate

In **Q1** and **Q2**, the derivative of a linear function will be a constant—the slope of the line. For the function $f(x) = bx + c$, $(f(x + h) - f(x))/h = (b(x + h) + c - (bx + c))/h = bh/h$. This expression equals b for all values of h other than 0.

In **Q3** through **Q7** the derivative approximation of the function $f(x) = ax^2$ will appear linear. For $a = 1$, the plot of $g(x) = (f(x + h) - f(x))/h$ will appear to equal $2x$. For $f(x) = ax^2$, $g(x) = (a(x + h)^2 - a(x)^2)/h = (a(x^2 + 2xh + h^2) - a(x)^2)/h = (2axh + ah^2)/h$. For $h \neq 0$, this expression equals $2ax + ah$. When h is near 0, the expression is close to $2ax$. For each value of a, the function g should approximate $y = 2x$, $y = x$, $y = 4x$, or $y = -x$. Putting together the results in **Q8**, the derivative function for $f(x) = ax^2 + bx + c$ is $y = 2ax + b$. (To explore whether derivatives actually add in this way, see Exploration 2.)

In **Q9**, changing the value of a in $f(x) = ax^2 + bx + c$ causes a vertical stretch or compression of the function plot, as well as changing the parabola's axis of symmetry. Varying b changes only the axis of symmetry—all the parabolas displayed are congruent. Varying c causes a vertical translation of the function. For the derivative function, changing a changes the slope of the linear function plot, and changing b causes a vertical translation of the function. Varying c has no effect on the derivative.

In **Q10** through **Q13**, the function g will appear quadratic. For $a = 1$, g will approximate $y = 3x^2$. For $f(x) = ax^3$, $g(x) = (a(x + h)^3 - a(x)^3)/h = (a(x^3 + 3x^2h + 3xh^2 + h^3) - a(x)^3)/h = (3ax^2h + 3axh^2 + h^3)/h$. For $h \neq 0$, this expression equals $3ax^2 + 3axh + h$. When h is near 0, the expression is close to $3ax^2$ (even though $3x$ is not a constant).

In **Q14**, the derivative function for f is $y = 3ax^2 + 2bx + c$. For **Q15**, the effects of changing the parameters a, b, c, and d on the plot of f can be explained using the idea of addition of function. When a is increased, for instance, the function plot appears more like the cubic ax^3, as the values from this term are much larger (or smaller, if x is negative) than the other terms. Similarly, increasing or decreasing b causes the plot to appear parabolic, at least locally near $x = 0$. For larger values of x, however, the plot still has the basic characteristics of a cubic, as the values of the cubic term are eventually much larger or smaller than the values of the squared term. There are many possible answers here.

Exploration 1

Students can predict the equations for the derivatives using graphical and numerical evidence. You may also wish to have the plot of the function g hidden at first to add another level to the exploration (see below.)

Exploration 2

Students can explore how combinations of functions are reflected in the derivative of the combined function. Only the first two questions are true, and students may be surprised how not true the second two questions are.

You can suggest the product relationship for derivatives using linear approximations. The tangent lines to two functions p_1 and p_2 at the point $x = a$ have the expressions $y = p_1'(a)(x - a) + p_1(a)$ and $y = p_2'(a)(x - a) + p_2(a)$. The product of these two approximations depends not only on the value of $p_1'(a)$ and $p_2'(a)$, but also the values $p_1(a)$ and $p_2(a)$, so value of the slope from $x = a$ to a point $x = a + h$ depends on them as well.

Extension

In Exploration 1, you may wish to have students first predict what the derivative approximation will look like. They can do this on their own, or with a secant line on the plot of f. To create a secant line, construct two points on the plot. Use an Animation button to animate the points at the same speed. Using the secant line, students can predict the plot of the derivative function before showing or plotting the function g.

Derivatives of Exponential Functions (page 70)

Prerequisites: Students should be familiar with the definition of the derivative, difference quotients as approximations, and the idea of transformations.

Sketchpad Proficiency: Intermediate. Students will need to measure coordinates, plot functions and points, and move objects in the plane. Beginners can start on the page Beginner's Start.

Activity Time: 60–75 minutes/40–50 minutes without the Explore More

Document: Exponent.gsp

General Notes: In this activity, students will examine graphs of the family of exponential functions $f(x) = a^x$, $a \geq 0$, and examine approximations of their derivatives. Their conclusions may suggest an expression for the derivative of $f(x) = 2^x$ and the entire family of exponential functions.

Sketch and Investigate

In **Q1,** where $0 < a < 1$, the graph of $f(x) = a^x$ is a decreasing function. Where $a > 1$, $f(x) = a^x$ is an increasing function. In **Q2,** where $a = 1$, $f(x) = 1$, so the function is constant. Where $a \leq 0$, a^x is not defined for all values of x. In **Q3,** the function plots both go through (0, 1). For $x > 0$, the greater a-value results in a steeper function and a steeper slope. For $x < 0$, the greater a-value results in a smaller value for the function and a shallower slope.

In **Q4,** the limit definition of the derivative of $f(x) = 2^x$ is $f'(x) = \lim_{h \to 0} \frac{2^{(x+h)} - 2^x}{h}$. Plotting the function $g(x) = \frac{a^{(x+h)} - a^x}{h}$ where $a = 2$ and h is small, as in **Q5** and **Q6,** results in a function that approximates the derivative. The values of this function at each value of x differ slightly from the actual derivative values; making h smaller creates a better approximation. The function g is also an exponential function. (See below.)

In **Q8,** $f(x) = a^x$ is constant for $a = 1$, so the derivative is 0. The derivative does not exist at any value of $x < 0$. In **Q9,** the derivative plot is negative and increasing as $f(x) = a^x$ is decreasing and concave up. As you increase a to values of greater than 1, the derivative plot gradually approaches the plot of $f(x) = a^x$ from below, as in **Q10.** In **Q11,** for $a \approx 2.7$, the graphs coincide, and for greater values of a the derivative plot is greater than $f(x) = a^x$.

Explore More

Finding that the derivative of $f(x) = a^x$ is a vertical stretch of itself for each value of a can lead to a rule for finding the derivative of the exponential function. For a single value of a, the ratio of the value of $f(x) = a^x$ to the value of its derivative is a constant; varying x does not change this constant. This constant c varies for each value of a. If $a = 1$, $c \approx 0$. If $a = 2$, $c \approx 0.69$. If $a = 3$, $c \approx 1.1$. Plotting the point (c, a) and constructing the locus of this point as a varies reveals the plot of a familiar function: $\ln(x)$. If students recognize that $c \approx \ln(a)$, or that the locus is a plot of $\ln(x)$, they may conjecture that $f'(x) = \ln(a)a^x$.

Extension

You may wish to add further support for the idea that the derivative of $f(x) = a^x$ is a vertical stretch of itself in order to support the idea that $f'(x) = \ln(a)a^x$. Recall that $g(x)$, an approximation of the derivative, is equal to $\frac{a^{(x+h)} - a^x}{h}$. This expression can be rewritten as $\frac{a^x a^h - a^x}{h}$, or $\frac{a^h - 1}{h}\left(a^x\right)$. In this form, the expression for the approximation of the derivative is a constant multiplied by a^x (because a and h are constants).

Can a function be its own derivative? For $a \approx 2.7$, $c = 1$, and the function plot and the derivative approximation coincide. Using the graphical evidence, students may conclude that $\frac{d}{dx}e^x = e^x$. You can further support this idea by examining the value of $\frac{a^h - 1}{h}$. For what value of a does this expression approach a value of 1 as $h \to 0$? An informal argument suggests that for $\frac{a^h - 1}{h} \approx 1$, $a \approx (h+1)^{1/h}$. Replacing h with $\frac{1}{x}$ gives the expression $\left(\frac{1}{x} + 1\right)^x$. The limit of this expression as x approaches infinity is one definition for the number e.

Derivatives and Transformations (page 74)

Prerequisites: Familiarity with the derivative as a function and with the effects of transformations of functions.

Sketchpad Proficiency: Intermediate

Activity Time: 50 minutes

Document: Transformations.gsp

General Notes: In this activity, students will investigate the effects of transformations on the plots of functions and compare the plots of derivatives of transformed functions with the plots of the original functions.

Sketch and Investigate

In **Q1**, changing the value of d results in a vertical translation; adding a constant to the function adds the same value (positive or negative) to the value of the function, so the plot of the function will be translated up or down by the value of the constant. In **Q2** through **Q5** the derivative is unchanged. Graphically, the differences between the values of the function at each value of x, and thus the rate of change of the function, do not change under a vertical translation. Symbolically, if $g(x) = f(x) + c$, then $g'(x) = f'(x)$.

In **Q6**, changing the value of c results in a horizontal translation. Adding a constant to x and then applying the function rule results in the same values of $f(x)$, but shifted by c. In **Q7** through **Q10** the derivative values are unchanged, but they occur at locations shifted by c as well. Graphically, the change in the function f over any interval will be the same if the interval is shifted left or right. Symbolically, if $g(x) = f'(x + c)$, then $g'(x) = f'(x + c)$.

In **Q11**, changing the value of a results in a vertical stretch or compression; multiplying the function by a results in y-values that are a times their previous value for each value of x. In **Q12**, if $a > 1$, the graph will stretch; if $0 < a < 1$ it will compress. In **Q13** through **Q16**, the derivative values are multiplied by the constant a, because the ratios between the points have been multiplied by a. If the values of a function at $x = 3$ and $x = 4$, for example, are 6 and 11 respectively, the values of the transformed function at the same two points are $6a$ and $11a$. The average rate of change between the two points, rather than being 5, is now $5a$. This argument can be extended to the instantaneous rate of change. Symbolically, if $g(x) = af(x)$ then $g'(x) = af'(x)$.

In step 18, changing the value of b results in a horizontal stretch or compression. Multiplying x by a constant and then applying the function rule results in the same values of $f(x)$, but stretched horizontally by a factor of $1/b$. Put another way, for the same y-value, the x-values of the function $f(bx)$ are $1/b$ times as large as the x-values of the function $f(x)$. (See the Explore More section for more on this point.) If $b > 1$, the graph will stretch; if $0 < b < 1$ it will compress.

For the same change in y the function g has a change in x only $1/b$ times as large, so the slope of any line between two points is b times as large. This argument can be extended to the derivative as well. The slope values, however, do not occur at the same values of x, because of the compression or stretch of the function plots. As a result, the derivative of $f(bx)$ at x is b times the derivative of $f(x)$ at bx.

For example, if the value of a function at $x = 3$ is 9, then the value of $f(2x)$ at $x = 1.5$ is also 9. If the value of a function at $x = 4$ is 13, then the value of $f(2x)$ at $x = 2$ is also 13. The average rate of change between the two points on the transformed function, rather than being $4/1$, is $4/0.5$, or 2 times the value for the original function. Symbolically, if $g(x) = f(bx)$, then $g'(x) = bf'(bx)$.

In **Q17** and **Q18**, $f'(x) = 6\cos(3x)$.

Explore More

These relationships can be explored using tangent lines. In **Q1**, the tangent lines at x and at $x + c$ will be parallel. The line moves left and right, and the slope stays the same. In **Q2**, the coordinates are $(x_P + c, f(x_P))$. In **Q3**, the equation for the tangent line is $y = f(x_P) + slope_f \cdot (x - (x_P + c))$. In **Q4**, the tangent line moves up and down as a changes. As the absolute value of a increases, the line gets steeper; as the absolute value decreases, it gets less steep. In **Q5**, the coordinates are $(x_P, af(x_P))$. In **Q6**, the lines match at $k = a$, and for $f(bx)$ they match at $k = b$.

Extension

You may wish to discuss why the value of P_g was chosen as it was. This choice allows the y-values of the points P_g and P to remain the same under the transformation caused by b.

By adjusting the slider for b, you can show how the slope of the function changes when the value of b is changed. Press the *Show Triangles* button on page 5 to illustrate the explanation given above.

Second Derivatives (page 78)

Prerequisites: Familiarity with the derivative as a function, and its properties and characteristics.

Sketchpad Proficiency: Beginner/Intermediate

Activity Time: 40–60 minutes

Document: SecondDerivative.gsp

General Notes: In this activity, students will explore properties and applications of the second derivative.

Sketch and Investigate

In **Q1**, if the second derivative is a constant, then the first derivative is a linear function. In **Q2**, if the constant is negative, then the function has negative slope. If the constant is positive, then it has positive slope. In **Q3**, if the second derivative is a constant, the original function is a parabola. This parabola opens downward if the constant is negative, and upward if the constant is positive. In **Q4**, with the second derivative an increasing or decreasing linear function, the first derivative is a parabola (concave up or concave down respectively). The larger the absolute value of b (the slope of the linear function), the greater the vertical stretch of the parabola. If the second derivative increases more quickly, then the slopes of the original function change more quickly.

In **Q5**, where $f''(x)$ crosses the x-axis from positive to negative, the first derivative has a maximum; where it crosses from negative to positive, the first derivative has a minimum. In **Q6**, where $f''(x)$ crosses the x-axis, $f(x)$ has a point of inflection. Given what happens to the first derivative, this is the point where the slope of f goes from increasing to decreasing or from decreasing to increasing. In **Q7** though **Q9**, if $f''(x)$ is always positive, the first derivative has a positive slope at all values of x (it is increasing on any interval), so the slope of f is always increasing. If $f''(x)$ is always negative, the first derivative has a negative slope at all values of x (it is decreasing on any interval), so the slope of f is always decreasing.

$f''(x)$	$f'(x)$	$f(x)$
Is positive	Is increasing	Is concave up
Is negative	Is decreasing	Is concave down
Is equal to 0	Has a maximum or minimum	Has a point of inflection

It is not possible to determine whether f is positive or negative from the second derivative, because the second derivative only gives information about whether the rate of change is increasing or decreasing, not whether it is positive or negative. (Recall that two functions can be vertical translations and have the same derivative; thus the second derivative can be negative, while the first derivative may be positive or negative.)

Using these relationships, it is possible to examine three graphs given only that one is f, another f', and a third f'' and determine which is which. One way to do this is to line up zeros with maximums and minimums to determine which of the functions is the derivative of another.

In **Q10** and **Q11**, the tangent line will (locally) lie above the plot of f when the second derivative is negative because the slope of the tangent line is decreasing. This occurs because the line is only an approximation of the function, and when the derivative is decreasing, the graph will fall below the tangent line both before and after the point of tangency. When the second derivative is positive, the tangent line will (locally) lie below the plot of f—the slope of the tangent line is increasing.

Exploration 1

One way to build the second derivative using just tangent lines and slopes is to create a point W on the plot of D_1. Measure its x-coordinate and then add h_2 to it. Evaluate D_1 at this new value, and use the values to plot a second point on the plot of D_1. Measure the slope of this line, and plot the x-coordinate of point W and the slope as a point in the plane. Then construct the locus of this point as W moves along the plot of D_1.

Explorations 2 and 3

These explorations test the ideas explored earlier in the activity. The connections between the values of the second derivative and the behavior of the first derivative are the key. Positive values of the second derivative cause positive slopes on the plot of the first derivative, which in turn either creates an increase in the original function (or a smaller decrease, depending on whether the first derivative is positive or negative itself).

In Exploration 3, it is possible for the position trace to continue to increase even when the acceleration is negative, because a negative value of the second derivative causes a decrease in the velocity, which decreases the slope of the position function but does not necessarily cause a decrease in position.

Newton's Method (page 82)

Prerequisites: Familiarity with the derivative, its characteristics, roots, and tangent lines to a graph.

Sketchpad Proficiency: Beginner/Intermediate

Activity Time: 30–50 minutes

Document: Newton.gsp

General Notes: In this activity, students will explore Newton's Method for solving equations and examine situations in which the method converges to a root and situations where it does not.

Sketch and Investigate

In **Q1**, $2^x = x^2$ is equivalent to $2^x - x^2 = 0$. In **Q2**, you can find a value of –0.76666 for the root of the equation. This solution produces values for 2^x and x^2 that are equal up to the thousandths digit. In **Q4** through **Q8**, when the seed is close to the root, Newton's Method will tend to converge to that root. What "close" means depends upon the function. With a seed between 1.5 and 1.52 for a different equation, $2^x = 2x$, Newton's Method will converge to the other root at $x = 1$ because of the location of the minimum of the function $2^x - 2x$. The locations of minimums and maximums will affect where the method converges because of its dependence on the slope of the graph.

If the method begins with a seed to the left of the leftmost root, it may or may not converge to that root. If the function has a maximum or a minimum at a point but does not intercept the x-axis near that point, the iterations of the method may bring you only very slowly to a root (or not at all if one of the iterations brings you to the x-value where the maximum or minimum occurs.)

In general, predicting what will happen with Newton's Method for different seed values is not easy, beyond the general principles described above. Near the x-values of the maximums or minimums points of a function plot, Newton's Method does not behave in a simple way, especially if the function has multiple maximums and minimums. See the Extension for a colorful way to explore what happens near these points.

Explore More

The question of whether Newton's Method converges can be further investigated using iteration. When you make the **Newton's Method** tool, point A must be split from the function plot so that the tool does not create new random points on the x-axis each time you use it.

Using the **Iterate** command allows you to construct a dynamic series of steps of Newton's Method. As you move point A along the function plot, you can see the value to which the method converges by measuring the x-coordinate of the terminal point of the iteration.

It is possible for Newton's Method to move farther and farther from a root, as in the example on page 3. For a function with a horizontal tangent and a root at the point where this tangent occurs, Newton's Method will not converge.

The equation for the first tangent line is $y = f'(x_0)(x - x_0) + f(x_0)$. The value of point B is then the solution to the equation $0 = f'(x_0)(x - x_0) + f(x_0)$, $x = x_0 + f(x_0)/f'(x_0)$. This value of x is then used as the new x_0. In general, the next iteration is found by the equation $x_{n+1} = x_n + f(x_n)/f'(x_n)$.

Extension

The values of x that converge to a particular root of a function when Newton's Method is used are called the *basins of attraction* for each of the roots. Some values of x do not converge to a root; this happens at the values of x at which the function has a horizontal tangent line. To visualize the basins of attraction, you can colorize point A using the x-value of the terminal point.

1. Measure the x-coordinate of the terminal point.
2. Select the seed point A and the measurement of the x-coordinate of your terminal point.
3. Select **Parametric** from the Color submenu in the Display menu. Set the range so that it encompasses all the roots of the function. Choose **Don't Repeat Color Range**.
4. Run tracing on for point A and animate point A along the x-axis.

Although at first glance you may see only three or four basins, the colors will exhibit chaotic behavior near the maximums or minimums of the function. Zooming in will reveal that there are smaller basins of each color that are not visible on a larger scale.

See the document **NewtonChaos.gsp** in the **Exploring Derivatives** folder for a demonstration of this chaotic behavior.

Following Tangent Lines (page 89)

Prerequisites: Familiarity with the derivative as a function, its characteristics, and the equation of a tangent line.

Sketchpad Proficiency: Beginner/Intermediate

Activity Time: 40–50 minutes

Document: FollowTangent.gsp

General Notes: Graphically, creating the antiderivative of a given function means creating a graph whose slope at any value of x is equal to the value of that function. In this activity, students will use the values of a given function to create a "tangent line" that is free to move around the plane, and use it to trace the antiderivative of a given function.

Sketch and Investigate

In **Q1**, the accuracy of the trace you create depends on how often you adjust to the changes in the direction in which the arrow is pointing. In order to create the exact plot of f, you must be able to make continuous changes, which is close to impossible! In **Q2**, the equation of the line through point R is $y = f'(x_R)(x - x_R) + f(x_R)$.

In **Q3**, moving point S to within 0.05 of R ensures that the tangent line value at S is less than 0.01 from the actual value of the function. The distance you must be from point S differs depending on what side of f you are on.

The construction of the tangent line from the coordinates of point R and the value of $f'(x_R)$ in steps 10 and 11 is an important part of this activity. When students draw an antiderivative themselves with paper and pencil, they must go through this same process, at least mentally. In **Q4**, when $f'(x)$ is positive, the tangent line will have a positive slope; when it is negative, it will have a negative slope. When $f'(x)$ is 0, the tangent line will be horizontal.

In **Q5**, moving point R vertically does not change the value of x_R, so it will not change the value of $f'(x_R)$, and the slope of the tangent line will not change. (Later, in **Q8**, this effect can be used to suggest that the antiderivative of a function is really a family of vertically translated functions.)

In **Q6**, wherever the slope of the tangent line is positive, the trace created by following the line will represent an increasing function; wherever the line has a negative slope, the function is decreasing. A local maximum or minimum will occur when the tangent line switches from positive to negative slope or vice versa. If the slope of the tangent line equals 0 but does not change sign, then the trace will have a point of inflection. For **Q7**, the answers to **Q6** should correspond to the answers to **Q4**.

Creating a button to dynamically move point R toward point Q will create a more exact antiderivative plot, because Sketchpad can change the direction in which point R moves virtually continuously.

In **Q8**, the traces created using a different initial location for point R will produce the same trace, vertically translated. The slope of the tangent line remains the same when you translate point R vertically, both at the initial location and throughout the movement of point R, because it is determined by the given function $f'(x)$. All of the different traces are antiderivatives, as vertical translations of a graph have the same slope and consequently the same derivative function.

Extension

To create your own questions, copy any of the pages 4–7 and edit the function $f'(x)$. Note that the given function is only labeled f'; it is not the derivative of a function f hidden on the page somewhere. When you create a derivative of f with Sketchpad's **Derivative** command, it will be labeled f', but you are unable to edit this expression.

If you would like to have the "answer" to the antiderivative question available on the page to be plotted and compared with the student's trace, create a function f, and find its derivative using the **Derivative** command.

To do this, calculate the derivative of f. Then double-click on the given expression for f' and delete it. Click on the computed f' expression, type x, and then close the parentheses. All of the objects in the sketch will now depend on this definition for f'. You can then plot the function f and create a button to hide it and the expression, so that after students make their trace, they can press this button to see how they did.

Step to the Antiderivative (page 93)

Prerequisites: Familiarity with the derivative and its properties and the equation of a tangent line; completion of "Following Tangent Lines."

Sketchpad Proficiency: Intermediate

Activity Time: 50 minutes

Document: StepTangent.gsp

General Notes: In this activity, students will use a custom tool to build an approximation of the antiderivative with a series of tangent line segments. In this way they will simulate the actions they might take when drawing an antiderivative on paper. Using the tool as a means of "guided practice," students sketch antiderivatives by hand and compare their drawings with the approximations created with the tool.

Sketch and Investigate

In **Q1**, students can check that the line shown has the needed properties by evaluating $f(x_R)$ and measuring the slope of line T. In **Q2**, the segment made with the tool, while still having a negative slope, has a slope closer to 0 than the previous one because the function f is negative but increasing. The value of f at x_Q is larger than the value of f at x_R. This should happen again for the next segment in **Q3**.

In **Q4**, the segments made will have a positive slope if f is positive, a negative slope if f is negative, and a zero slope if f is 0. With paper and pencil, tracing from the point at which you are about to make a segment to the plot of f can help you "line up" the segment you will draw with the behavior of f and make the correct decision about how to draw the next segment. In **Q5**, for instance, if you see that the value of f has increased since you made your last segment, you know that the slope will be greater; if f has decreased, the slope will be less.

Note that "greater" means closer to 0 if the slope is currently negative, but farther from 0 if the slope is positive. Using the tool in this activity can help students make the right decision when sketching with paper and pencil if they observe this connection between the slope of the segment constructed by the tool and the value of f.

In **Q6**, students may be concerned that their paper and pencil sketch of the antiderivative has too great or too little of a range. You may wish to have them focus on whether their drawing has maximums or minimums at the correct values of x. For the purposes of this activity, it is not necessary to draw an exact antiderivative. Students can compare their drawings with the one created in the document to see where they have inaccuracies and why.

Printing the page with the grid showing can be an aid to drawing, as students can then read the values of the function and create tangent segments with greater accuracy. However, in **Q9**, students should also know that it is possible to draw an antiderivative without a scale on the y-axis by judging the relative change in the values of f. In the sketch, you can hide the y-axis and construct a line between the points (0, 0) and (0, 1) to create an axis without a scale, as was done on page 6.

In **Q7**, the series of segments is a more accurate approximation of the antiderivative because the tangent line is followed for a shorter distance, minimizing the error introduced with this approximation technique. In **Q8**, the traces created by using a different initial location for point R produce the same trace, vertically translated. In **Q9**, adding a constant to a function does not change its derivative, so the antiderivatives of a given function are a family of vertically translated graphs.

In **Q10** through **Q12**, where $f(x)$ is positive, the antiderivative approximation is increasing—the segments have a positive slope. Where $f(x)$ is negative, the antiderivative approximation is decreasing—the segments have a negative slope. Where $f(x)$ is increasing, the antiderivative approximation is concave up—the slope of the segments is increasing. Where $f(x)$ is decreasing, the antiderivative approximation is concave down—the slope of the segments is decreasing. The antiderivative approximation has a maximum where $f(x)$ changes from positive to 0 to negative, and a minimum where $f(x)$ changes from negative to 0 to positive.

Explore More

Students create a tool that can be used on graphs that are not Sketchpad function plots. With this tool, you must determine the distance that you follow the tangent line by clicking on points on the "graph," unlike the tool used earlier, where this distance was determined by h. In **Q2**, at $x = -2$, the antiderivative changes from a zero slope to a positive slope; at $x = 2$, it changes from a positive to a negative slope. At $x = 4$ the antiderivative has a point of inflection. At $x = 6$, the antiderivative has a positive slope at its endpoint.

Plotting the Antiderivative (page 97)

Prerequisites: Familiarity with antiderivatives from a graphical perspective—construction from a tangent line. The previous two activities give very helpful background. Knowledge of derivative expressions for elementary functions is important since predicting antiderivatives from approximations is worked on in this activity.

Sketchpad Proficiency: Beginner/Intermediate. Plotting points, plotting functions, measuring coordinates, working with iteration.

Activity Time: 50 minutes

Document: Antiderivatives.gsp

General Notes: In this activity, students will use Euler's method to numerically and graphically construct an approximation of the antiderivative of a given function. Students will experiment with a dynamic approximation of the antiderivative of a given function to explore properties of antiderivatives.

Sketch and Investigate

In **Q1**, the function f determines the slope of the tangent line through the point R, so the coordinates of point Q will be $(x_R + h, y_R + f(x_R) \cdot h)$. In **Q2**, the antiderivative is a vertical transformation of the negative cosine function. Have students plot a cosine function that passes through the point (–2, 1) and see how it compares with this approximation.

In **Q3**, the coordinates after 40 iterations with step size 0.1 are (2, 0.90907). In **Q4**, the coordinates after 80 iterations with step size 0.05 are (2, 0.95454). As you increase the number of iterations, the y-value of the terminal point will approach 1 (as long as you increase the number of iterations so that the approximation reaches $x = 2$). By using the initial conditions, and knowing that the antiderivative of $\sin(x)$ is $-\cos(x) + C$, you can show that the iterations approximate the function $F(x) = -\cos(x) + 1 + \cos(-2)$. In **Q5**, using a larger number of iterations will make the iterated image of discrete points appear more like a smooth plot.

In **Q6**, Euler's method with 80 iterations gives $F(2) \approx 3.06429$. In **Q7**, the approximation is a vertical translation of a cubic function. If students have learned expressions for derivatives, they may determine that the function being approximated is $(1/3)x^3 + C$.

Exploration 1

The antiderivatives for each page are:

3. $F(x) = 0.5x^2 + bx + c$
4. $F(x) = x^3 - x^2 + x + c$
5. $F(x) = -\cos(x) + 0.25x^2 + c$
6. $F(x) = \ln|x|/2^x + c$
7. $F(x) = \ln|x| + c$

In **Q1**, the approximation is 1.46172. To reach $x = 5$, a total of 299 iterations are needed, because the first point plotted counts as the first iteration.

Exploration 2

In **Q1,** the approximation of the antiderivative of $f(x) = \sin(x) + c$ closely approximates a line as c is increased or decreased. If c is adjusted so that f is always positive, then the approximation increases on all intervals. A similar result holds if c is adjusted so that f is always negative. Wherever f is negative, the approximation decreases; wherever it is positive, the approximation increases. In **Q2**, the antiderivatives of $f(x)$ and $f(x) + c$ differ by the addition of a linear term, cx. In other words, if $g(x) = f(x) + c$ and $F(x) + C_1$ is the antiderivative of $f(x)$, then the antiderivative of $g(x)$ is $G(x) = F(x) + cx + C_2$, where C_1 and C_2 are arbitrary constants.

In **Q3**, the antiderivatives of horizontally translated functions are not usually horizontal translations of each other; the initial conditions matter. Symbolically, if $g(x) = f(x + c)$ and $F(x) + C_1$ is the antiderivative of $f(x)$, then the antiderivative of $g(x)$ is $G(x) = F(x + c) + C_2$, where C_1 and C_2 are arbitrary constants. So G is a horizontal and vertical translation of F.

In **Q4**, the antiderivatives of vertically compressed or stretched functions will not be vertically compressed or stretched transformations of each other, unless the initial condition is (0, 0). For other initial conditions, the antiderivative of $cf(x)$ will be $G(x) = cF(x) + C_2$, where $F(x) + C_1$ is the antiderivative of $f(x)$.

In **Q5**, when you horizontally compress the function f, its antiderivative will be horizontally *and* vertically compressed by the same factor. In other words, if $g(x) = f(c \cdot x)$ and $F(x) + C_1$ is the antiderivative of $f(x)$, then the antiderivative of $g(x)$ is $G(x) = (1/c) \cdot F(c \cdot x) + C_2$, where C_1 and C_2 are arbitrary constants.

A Field of Slopes (page 101)

Prerequisites: Familiarity with antiderivatives from a graphical perspective. Knowledge of tangent lines and approximating with tangent lines a must! The previous three activities are recommended but not necessary.

Sketchpad Proficiency: Intermediate. Quite a few constructions.

Activity Time: 50 minutes

Document: Slopefield.gsp

General Notes: In this activity, you will create slope fields with a *tracer*—a small segment whose slope is determined by a given equation. This segment will serve as your guide as you learn how to draw slope fields and explore the relationship between a differential equation and its slope field.

Sketch and Investigate

In **Q1** and **Q2**, the trace of the segment will be close to horizontal when the y-value of f is closest to 0 or where f has a root. The trace of the segment will be closest to vertical where f has a maximum or minimum. You can find the exact slope of any segment by finding the y-value of the function at that x-value in the slope field. Where the differential equation is strictly a function of x, all the segments in a column will have the same slope. In **Q3**, since the slope field pictured consists of positive segments with identical slopes, $f(x)$ must be constant, and in this case, positive.

Segment RQ is constructed to have a slope equal to $f(x_R)$; you can follow this segment to reveal the graph of F, a particular solution to the differential equation. This will give you a sense of how to sketch a solution curve with paper and pencil.

In **Q4**, you can substitute the values of –9 and 7 into the function $F(x) = -\cos(x) + C$, to arrive at the equation $F(x) = -\cos(x) + 7 + \cos(-9)$. In **Q5**, any other initial point will produce a vertical translation of the function $F(x) = -\cos(x) + C$.

In **Q6** through **Q8**, the segments in the slope field will have positive slopes where $x > 0$, and negative slopes where $x < 0$. At any value of x, the slope of the segments in the field will be x, so the slopes of the segments will increase as you move away from 0 in the positive direction, and decrease as you move away in the negative direction. Because $f(x) = x$, the slope field will define the set of parabolas $\{y = 1/2x^2 + C \mid C \text{ is any real number}\}$. (See Q7 on page 109 for the slope field.)

In **Q9–Q12**, the slope field has segments with a slope close to 0 near the x-axis, with positive and increasing slopes as you move up in the coordinate plane and negative and decreasing slopes as you move down. The particular solutions are exponential functions. The actual equation for the solutions is $y = e^{x+C}$, or Ae^x, because $y' = Ae^x = y$. Or in words, the slope of this function is equal to the y-value of the function at any point. $y = 0$ is also a solution, which the slope field suggests. (See Q9 on page 109 for the slope field.)

Explore More

The slope field for $dy/dx = 0.5y$ is similar to the one for $dy/dx = y$, in that all the segments on any horizontal line are the same. On each horizontal line, however, the slopes are only half what they were for $dy/dx = y$. The solution to the differential equation $dy/dx = ky$ is $y = Ce^{kx}$, which you can show by taking the derivative of this function. The derivative of y is $Cke^{kx} = k(Ce^{kx}) = ky$. The function $y = 0$ is also a solution, as before.

The slope field for $dy/dx = y - 2$ is similar to the one for on $dy/dx = y$, in that all the segments on any horizontal line are the same. The entire field will be shifted vertically by 2. This suggests that the solution to the differential equation is $y = Ae^x + 2$, which you can confirm by taking the derivative. We have $y' = Ae^x + 0 = Ae^x$ and $y - 2 = (Ae^x + 2) - 2 = Ae^x$. So $y' = y - 2$. The equation $y = 2$ is also a solution, as the slope field suggests.

Extension

You can also construct a slope field on page 2 with an additional locus—besides the locus of grid points. In this way, the slope field will update when you edit the expression *SlopeAtR*. With point R merged to the locus of grid points, select point R and the segment and choose **Construct | Locus**. You may need to increase the number of samples in the locus to see the whole field.

For any differential equation whose denominator never equals 0, you can split point R from the locus and press the *Move R→Q* button to create a trace of a particular solution. Note that the trace of the solution will only go in one direction, and will behave incorrectly if the slope field has vertical segments. For a more flexible slope field and tracer, see the next activity.

Stepping Through the Field (page 105)

Prerequisites: Familiarity with the concept of antiderivatives and slope fields for a differential equations. Completion of "A Field of Slopes" is recommended.

Sketchpad Proficiency: Beginner/Intermediate

Activity Time: 50 minutes

Document: Euler.gsp

General Notes: In this activity, students will experiment with slope fields for a variety of differential equations by creating them with a trace as in the previous activity or by examining a completed slope field. They will also visualize the numerical algorithm known as Euler's method to approximate solutions to differential equations.

Sketch and Investigate

In **Q1** and **Q2**, the closer the point is to the lines $y = x$ or $y = -x$, the closer the segment in slope field will be to a slope of 1 or –1. Toward the x-axis, the segments will have more of a vertical slope, and toward the y-axis, more of a horizontal slope. In **Q3**, the solution curve has the shape of a hyperbola. In **Q4**, apply Euler's method to the point (–4.8, 3.75) by substituting these coordinates into the expression $y_R + slope \cdot h$. This expression calculates the change in y from point R over a change in x of h for the given value of *slope*. The result is an approximation of the value of the solution at a point h from the initial point. The calculation is then $3.75 + (-4.8/3.75)(0.2) = 3.494$. So the next approximated point is (4.6, 3.494).

In **Q5**, the 12th iteration of Euler's method will produce a point far from the previous points. This is due to the large slope value that the differential equation produces in the previous iteration. This happens because the value of y_R at the 11th iteration is close to 0: $y_R = 0.08$, and y_R is the denominator of the differential equation.

In **Q6**, the coordinates (–3, 0.643) represent the estimated point on the solution curve for the differential equation, or $F(-3) \approx 0.643$. In **Q7**, move point R to the point (1, 0.5) and set h to 0.1. From there, you will need 20 iterations (counting point Q—so 19 on the iteration panel) to reach $x = 3$. Make sure that your direction control is set to the right. Otherwise, your approximation will yield a point on the x-axis, and in this case, an undefined slope. So here, $F(3) \approx 2.8889$.

Exploration 1

The answers for **Q1** are on pages 109 and 110. These plots can be copied for **Q2** if students don't have access to a printer. Particular solutions found in **Q3** and descriptions in **Q4** will vary.

Exploration 2

The differential equation $(x^2 - 4)/(x + y)$ has value of 0/0 at the point (–2, 2). At any other point along the line $x = -2$, the slope field is horizontal, and at points to the left and right, the field is generally vertical; the segments near the point "lead" to the point (–2, 2). Note that at the point (2, –2), although the value of the differential equation is also 0/0, the nearby segments lead away from the point.

To avoid the problem of large jumps from one iterated point to another, the iterated image on page 6 of the document was created using a constant value for the length of each jump, rather than a constant value of h.

Extension

Find a text that discusses predator/prey population relationships. Such relationships are often modeled with differential equations. Use one of the pages of the document to visualize the relative populations of two species over time.

Investigate the logistic curve and its applications. Use one of the pages of the document to model the situation.

Building Area (page 113)

Prerequisites: Familiarity with the concepts in "Going the Distance" and "Areas and Limits" is helpful but not required.

Sketchpad Proficiency: Beginner/Intermediate

Activity Time: 50 minutes

Document: BuildArea.gsp

General Notes: In this activity, students will approximate the area between a function plot and the x-axis by building rectangles and trapezoids with custom tools and investigate the accuracy of each approximation.

Caution: The exact area calculation on pages 3–4 of the document is only valid for functions in the form $f(x) = a(x - b)^2 + c$.

Sketch and Investigate

In **Q1**, when point N is on the left, the height of the rectangles is determined by the value of the function on the left side of each of the subdivisions from a to b. When point N is on the right, the right side of each subdivision is used instead. In **Q2**, as the value of n increases, the widths of the rectangles decrease and the error decreases. For the given function, the area approximation approaches a limit of 52 1/3.

In **Q3**, the area measurements can be negative since the value of the function can be negative. Multiplying the positive value of h by the negative value of $f(x)$ results in a negative area calculation. The sum of the rectangles can be negative if the negative area calculations sum to more than the positive area calculations. In **Q4**, multiply the value of h by the height of each of the rectangles. The height of the rectangles is the value of the function at each of the left endpoints of the subdivisions.

In **Q5**, as described above, the height of a left sum rectangle is determined by the value of the function on the left side of the subdivision on which it is constructed. The height of a right sum rectangle is determined by the value of the function on the right side of the subdivision. In **Q6** through **Q9**, whether the right sum or the left sum is an overestimate or an underestimate will depend on the function and the locations of point A and B.

Explore More

In **Q1** through **Q5**, the formula for the area of the trapezoids made with the tool is $h \cdot (f(x_L) + f(x_R))/2$, where x_L and x_R are the values of x on the left and right side of the subdivision on which the trapezoid is built. For the midpoint rectangles, the formula is $h \cdot f((x_L + x_R)/2)$. Using 10 trapezoids, the percentage error is approximately 0.8%. The 10 midpoint rectangles will produce an error of only 0.4%. The 10 right rectangles will produce an error of 3.3%. In this case, the midpoint rectangles produce the least error.

Extension

The tools used in this activity can be constructed using the following steps:

1. Follow steps 1–7 in the activity. Note that n could be a parameter rather than a value calculated using a slider.
2. Construct an independent point I and measure its x-coordinate.
3. Add the measurement h to x_I.
4. Measure the y-coordinate of point I.
5. Plot the point $(x_I + h, y_I)$. Do not label.
6. Next, to make a left rectangle tool, plot the points $(x_I, f(x_I))$ and $(x_I + h, f(x_I))$. To make a right rectangle tool, plot the points $(x_I, f(x_I + h))$ and $(x_I + h, f(x_I + h))$.
7. In either case, select the four points that form the rectangle and construct the quadrilateral interior. You can also construct segments connecting the four points. (See the note below if you want your tool to measure area as well.)
8. Select points I and $(x_I + h, y_I)$, the function f (not the plot), h, the quadrilateral interior, and the segments. Select **Create New Tool** from **Custom** tools. Give your tool a name.

To automatch f and h, open the Script View. Double-click on Given function f and check Automatically Match Sketch Object. Click OK and then do the same with Given measurement h.

Note: To have an area calculation also appear with the tool, calculate the area of the rectangle by multiplying the h measurement and the $f(x)$ calculation that determines the height of the rectangle. Relabel this calculation *area* and select Use Label in Custom Tools. Select this calculation in addition to the other objects when creating the tool.

For other variations on area tools, see **ToolTips.doc** as well as the next activity, "The Trapezoid Tool."

The Trapezoid Tool (page 117)

Prerequisites: Students will use the geometric formula for the area of a trapezoid.

Sketchpad Proficiency: Beginner. Students will need to measure coordinates and move objects in the plane.

Activity Time: 40–50 minutes

Document: TrapezoidTool.gsp

General Notes: In this activity, students will build a custom tool to construct trapezoids between a function plot and the x-axis and use this tool to make conjectures.

Sketch and Investigate

In **Q1**, students should find a total area of about 0.688 if they use the x-values of 0, 0.5, 1, 1.5, and 2. They will get various other answers for different x-values. The actual area, in **Q2**, is 2/3. The x-values that give the smallest percentage error are 0, 0.5, 1, 1.5, and 2—equally spaced points with interval width (2 – 0)/4, or 0.5.

Exploration 1

In **Q1**, students should find a total area of about 2/3 but their result will depend on where they place their trapezoids. In **Q2**, to achieve the minimum percentage error, the points should be equally spaced starting at 0 with width (2 – 0)/8, or 0.25.

Exploration 2

In **Q1**, using the **Trapezoid** tool on a curve that goes below the x-axis will result in a negative area calculation, since the calculation is based upon the values of the function.

In **Q2**, using the tool on f at one point where f is below the curve ($f < 0$) and the other point where f is above the curve ($f > 0$) results in a "trapezoid" that will appear as two triangles. You can confirm that the sum of the area of these two triangles, one of which is a negative area and the other positive, is correctly calculated with the trapezoid formula.

Answers to **Q3** may vary, but the total area should be –8.71681. If point B is placed precisely at $x = -2$, then the area will be $-15 + 2\pi$.

Extension

Sometimes you may be interested in finding an approximation for the integral of a function that is not defined by an algebraic expression, but by a set of data points. In addition to calculating the approximation, it can be instructive to visualize the calculation by drawing the trapezoids or rectangles.

When a function is defined by a set of data points, the tool constructed in this activity cannot be used, as it requires selecting a function as one of the givens. Suppose, for instance, that you wanted to find the trapezoidal approximation for the integral of a function such as the one in the table below.

t (sec)	0	5	10	15	20	25
$v(t)$ (ft/sec)	12	16	24	25	20	18

After plotting the points, you could create a new trapezoid tool in the following manner:

1. Measure the y-coordinates of two of your data points, A and B.
2. Measure the x-coordinates of points A and B, and then calculate $x_A - x_A$. Relabel this calculation *zero*.
3. Plot (x_A, *zero*) and (x_B, *zero*). Construct the interior of the trapezoid formed by your four points.
4. Calculate the area of the trapezoid using the expression $0.5 \cdot (y_A + y_B) \cdot (x_B - x_A)$. Relabel this calculation *AreaT*, and select the Show in Custom Tools box.
5. To make the tool, choose the objects that are necessary to construct your final goal and then the objects you want to end up with. So here, you need the points A and B, and then the region's interior and the area measurement. (You may also want to select the points on the x-axis, but that is optional.)
6. Choose **Create New Tool** from **Custom** tools and give it a name.

You can use this tool on any data points, and it will work just like the tool in the activity. You can also use it on a function plot; all you need to do is click on two locations on the plot.

You can also easily make a tool to create left or right sum rectangles as well. Use the method described above, but then construct a point at (x_B, y_A) for a left sum, or (x_A, y_B) for a right sum. Use your points to construct the interior of the rectangular region, and then compute the area $(y_A) \cdot (x_B - x_A)$ for the left sum, or $(y_B) \cdot (x_B - x_A)$ for the right sum. See the document **Areatools.gsp** for some completed examples.

Accumulating Area (page 120)

Prerequisites: Students should be familiar with graphs of functions and area formulas.

Sketchpad Proficiency: Beginner/Intermediate. Students will measure coordinates and move objects in the plane.

Activity Time: 30–50 minutes

Document: Area2.gsp

General Notes: In this activity, students will informally extend the concept of area using the notation of integrals with a function consisting of geometric figures.

Sketch and Investigate

In **Q1**, students should find that the value of the integral is close to $6 + \pi$. The difference lies in the exact locations of the points *start* and *finish*. You can create *Move* buttons to move these points to the exact locations. If these points are moved in this fashion, the *Calculate Area* button will give you a result of 9.14159. In **Q2**, the value of $\int_{-2}^{4} f(x)\,dx$ is $6 + 2\pi$. In **Q3**, answers will vary. Using the **Area** tools will reveal that the value is the opposite of the previous answer, namely $-6 - 2\pi$. One way to motivate this property of integrals is to appeal to the way area was calculated in the previous activity, "The Trapezoid Tool." In **Q4**, the formula used to calculate the area introduces a negative sign when the "finish" point comes after the "start" point.

In **Q5** through **Q7**, the integral consists of two triangles; they are congruent, so the integral will equal 0. In **Q8**, the value of $\int_{-6}^{-3} f(x)\,dx$ is –7.5.

In **Q9**, the notion of the area function is introduced, and the area from –6 to –6 is 0. In **Q10**, $A(-3)$, which equals $\int_{-6}^{-3} f(x)\,dx$, is –7.5, as above.

In **Q11** through **Q14** the area trace decreases immediately because the value of the area is negative. An area of –8 will accumulate first, so the area plot will not simply become positive at $x = -2$, but it will increase because the area between the function plot and the x-axis becomes positive at this point. At the places where the area trace crosses the x-axis, there is an accumulated area of 0—the integral from –6 to that point is 0. At the points where the function f crosses the x-axis, the area trace has a maximum or a minimum; it is at these locations that there is a change from adding positive area to adding negative, or vice versa.

Answers to **Q15** may vary. Students may notice that when the original function is positive, the area function increases, and when it is negative, the area function decreases. They may also notice that when the original function is increasing, the area function is concave up, and when it is decreasing, the area function is concave down.

Explore More

In **Q1**, the area function moves vertically as point *start* is moved. In **Q2**, the area plot will move up when the original function is positive, and down when the original function is negative.

Note that the area function does not change in any other way—if the original function is unchanged, the same areas are involved—but the starting point causes some of them to be counted as negative and some of them positive. In **Q3**, in integral notation, $\int_a^x f(t)\,dt = \int_b^x f(t)\,dt + C$, where C is $\int_a^b f(t)\,dt$.

In **Q4**, although the function itself may have a positive value, point I, which plots the area, may have a positive or negative value, depending on where the area accumulation began and the function's values to the left of the current location of point I.

Extension

Students may be interested in how the **Area** tools in the sketch used in this activity compute the area and whether this area is "exact." Here, the area calculation comes from the value of the antiderivative of the original function (in this case, it is a piecewise function).

To create a similar demonstration, you can plot a function and then (on your own) determine its antiderivative. Build this function in the calculator, and label it $g(x)$. Then construct a point on the x-axis. (This is your point *start*, as in the activity.) Measure its x-coordinate, x_{start}. Calculate $g(x_{start})$. Then create a new function $g(x) - g(x_{start})$. ($g(x_{start})$ is a constant.) Label this function $A(x)$. Next, create another point on the x-axis. (This is your point P, as in the activity.) Find the value of $A(x_P)$. This value gives you the area calculation that you need; it is 0 when $x = x_{start}$ and yields the area for other values of x as you move point P along the axis. You can then plot the area calculation as you did in the activity.

Area and Integrals (page 124)

Prerequisites: Familiarity with approximating the area of regions using rectangles and trapezoids. Familiarity with the concepts explored in the last three activities is helpful, but not required.

Sketchpad Proficiency: Beginner/Intermediate

Activity Time: 45 minutes

Document: AreaIntegral.gsp

General Notes: In this activity, students will visualize a definition of the integral by constructing rectangles with a custom tool and exploring a dynamic approximation of the integral. Students will use their explorations to speculate about properties of the integral.

Sketch and Investigate

In **Q1**, student's answers will vary, as they will form the partition of the interval on their own. The lower sum, however, will be an underestimate of the integral from a to b. In **Q2**, the upper sum will be an overestimate of the area. The difference between these values may be large because of the differences in the y-values at the sample points. In **Q3**, the difference between the upper and lower sum could be decreased by using more rectangles.

Caution: The sums on pages 3 and 4 will eventually become over- or underestimates as n increases, depending on the number of maximums and minimums in the interval. In **Q4–Q6**, for $n = 6$, the overestimate has a value of 57.61, the under-estimate, a value of 47.98. For $n = 12$, the over-estimate has a value of 54.86; the underestimate, a value of 50.04. For $n = 24$, the overestimate has a value of 53.57; the underestimate, a value of 51.15. For $n = 48$, the overestimate has a value of 52.94; the underestimate, a value of 51.74. For $n = 96$, the overestimate has a value of 52.64; the under-estimate, a value of 52.03. For $n = 396$, the over-estimate has a value of 52.41; the underestimate, a value of 52.26. Each of the sums is converging to the same number, a number close to 52.3. The actual value of the integral is 52 1/3.

In **Q7**, the underestimate and overestimate approach 7.08 as n increases. In **Q8**, the area of one rectangle is enough to find the value of the integral from –2 to 3, which for $f(x) = 3$ is 15. In **Q9**, the **Riemann Rectangle** tool will calculate $f(x_i)h$, with h found by subtracting the value of the second point matched from the first point matched. When you create the rectangle with the tool, matching the second point whose x-coordinate is less than that of the first point matched will cause h to be negative.

Because the area depends on the value of h and the value of $f(x_i)$, this area can be negative if the value of h is negative. In **Q10**, $\int_{-2}^{3} f(x)dx = -15$. In **Q11**, $\int_{-2}^{3} f(x)dx = -15$, and $\int_{3}^{-2} f(x)dx = 15$. The value of the function is negative, but because the integral is evaluated from a larger value to a smaller value, the integral is positive.

In **Q12** through **Q15**, reasoning from the definition of the integral as a limit of a sum of rectangles, $\int_{-3}^{-1} f(x)dx$ is a limit of a sum of rectangles with positive area, so it has a positive value. Similarly, the integral $\int_{-1}^{1} f(x)dx$ has a negative value. The integral $\int_{-3.5}^{-1} f(x)dx$ probably has a positive value, because the limit of the sums of the rectangles with negative area from –3.5 to 3 probably has a smaller absolute value than that of the positive rectangles from –3 to –1. The integral $\int_{-3}^{1} f(x)dx$ will have a value of 0 because the limit of the sum of the positive rectangles from –3 to –1 will have the same absolute value as that of the negative rectangles from –1 to 1. In **Q16** the integral from –3.5 to 1.5 would also have a value of 0, assuming the function is symmetric about the line $x = -1$.

Explore More

Students may guess that the suggested properties are true after thinking about the definition of the integral as the limit of a sum of rectangles. For **Q1** and **Q2**, they can experiment on page 5 by adjusting the slider for d and observing the change in the sum of the upper or lower sum.

Extension

This activity, "Area and Limits," and the three previous activities, form a set of activities involving approximation of integrals and the connections between area and integrals. In "The Area Function," students begin to explore the idea that a function can be used to find the area of a particular region or the value of a particular integral.

The dynamic Riemann sums used in the document **AreaIntegral.gsp** and the dynamic left and right sums in **BuildArea.gsp** were constructed using iterated images of a rectangle. The value of n defined by a slider allows the iteration to be constructed to a specified depth. For more on iteration, see *The Geometer's Sketchpad Reference Manual* or Sketchpad's help system.

The Area Function (page 128)

Prerequisites: Familiarity with calculating the area between the plot of a function and the x-axis on an interval using geometric objects.

Sketchpad Proficiency: Beginner/Intermediate. Plotting points and functions.

Activity Time: 50 minutes

Document: AreaFunction.gsp

General Notes: In this activity, students will explore the concept of the integral as a function that accumulates area from a fixed value of x to a variable of x. Students will build plots of $\int_a^x f(t)dt$ for some different functions and become familiar with the idea of the integral as a function.

Sketch and Investigate

In **Q1**, the function f is a constant, so the area accumulates at a constant rate from $x = -4$ to the variable point $x = x_P$ and the region between the plot of the function and the x-axis is a rectangle. The trace does pass the vertical line test for functions. Although the rectangular region is always above the x-axis, the area function is negative left of $x = a$. This is because the definition of the integral is dependent on the widths and heights of the rectangles. Any rectangle formed to the left of $x = a$ in this example will have a *negative* width because if $x_P < a$ then the width equals $x_P - a$, which is less than 0. Multiplying by a positive height gives a negative signed area.

In **Q2**, $A(x_P)$ is always increasing; it is zero at $x_P = a$. In **Q3**, the plot of $A(x_P)$ is a linear function. The equation is $A(x_P) = 0 + 3(x_P + 4)$ using point-slope form and the plot, which can be shown by multiplying the height, 3, by the width, $(x_P - (-4))$.

The area of a trapezoid is $0.5(x_P - a)(f(x_P) + f(a))$ for step 11. The problem is what happens if $x_P > 0$ and there are really two triangles instead of a trapezoid. The sum of the area of the two triangles is $0.5x_P \cdot f(x_P) + 0.5(0 - a) \cdot f(a)$. The two formulas are equivalent because $0.5(x - a)(f(x) + f(a))$
$= 0.5(x - a)(2x + 2a) = 0.5(2)(x - a)(x + a)$
$= 0.5(2)(x^2 - a^2) = 0.5(2x)(x - 0.5)(2a)a$
$= 0.5x \cdot f(x) + 0.5(-a) \cdot f(a)$.

In **Q4–Q7**, the function $A(x_P)$ is decreasing on the interval $(-\infty, 0)$, increasing on the interval $(0, \infty)$, and has a minimum where $f(x) = 0$ at $x_P = 0$. The trace looks like a parabola. This can be proven by using the points $(-4, 2(-4))$ and $(x, 2x)$ in the formula for the area of the trapezoid:

$0.5(x - a)(f(x) + f(a)) = 0.5(x + 4)(2x - 8) = x^2 - 16$. Substituting any value into this expression gives the integral from -4 to your value.

In steps 14–23, students use an approximation of the integral to create a plot of accumulated area.

The idea behind these steps arises from a Riemann sum. You can create a plot of accumulated area from the area of each of the rectangles by plotting a sequence of partial sums of area. For example, imagine a right Riemann sum with 8 subdivisions. At x_0, no area has been accumulated. At $x = x_1$, the area of the first rectangle, $f(x_1)h$, has been accumulated. At $x = x_2$, a total area of $f(x_1)h + f(x_2)h$ has been accumulated. Each of these partial sums can be plotted in the plane at the corresponding value of x up to $x = x_i$.

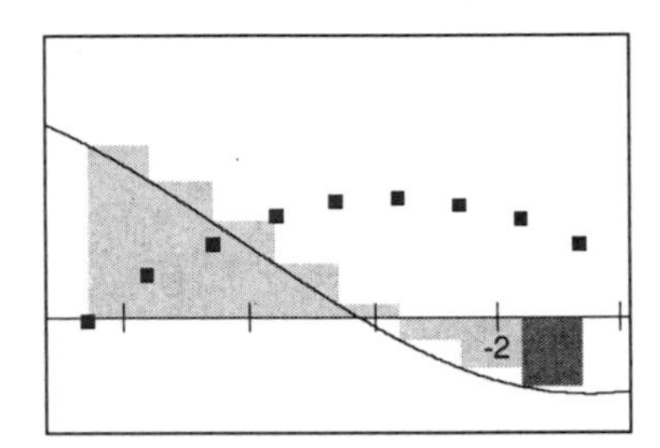

The greater the number of subdivisions used, the more accurate the approximation of the integral function $A(x)$ at any value of x.

In **Q8**, a plot created in this manner shows an overestimate when the trace of the rectangle goes above the curve and an underestimate, below.

Explore More

In **Q1**, the antiderivative is $\ln(x)$; in **Q2**, it is $x^3/3$; and in **Q3**, it is $e^x - 1$.

Extension

As suggested in the Explore More section, this activity can be used to help students think about the integral as an accumulator of the rate of change of a function. You can relabel the axes to reflect different contexts and then discuss the meaning of the area function in each context.

For example, consider a graph that gives the rate of water flowing out of a pipe in gallons per hour as a function of time. What would the area function for this graph represent?

Or, consider a graph of the cost per mile of drilling underground as a function of the distance underground. What would the area function for this graph represent?

Plotting the Integral (page 132)

Prerequisites: Familiarity with the integral as a function and properties of integrals. Completion of "The Area Function" is highly recommended.

Sketchpad Proficiency: Beginner/Intermediate

Activity Time: 50 minutes

Document: PlotIntegral.gsp

General Notes: In this activity, students will explore the connections between the plots of f and $\int_a^x f$ and learn how to plot the functions defined by integrals.

Sketch and Investigate

In **Q1**, the function $A(x_P)$ is always increasing, and has a value of 0 at $x = a$. In **Q2**, moving point P moves point A_P along the locus. When you change the value of a, the locus itself will move so that the location of the zero remains with $x = a$. In **Q3**, the expression for $A(x_P)$ is $3(x_P - a)$, because the height of the rectangle is 3, and the width is $(x_P - a)$. The locus is a plot of the function $3(x - a)$. In **Q4**, adjusting the slider allows you to see the plot of the family of functions $f(x) = c$. As you vary c, the function plot will move vertically. The locus for $A(x_P)$, which is a linear function, will change slope. Using the formula for the area of a rectangle as in Q3, $A(x_P) = c(x_P - a)$. The expression for the locus is then $c(x - a)$.

In **Q5**, the function $A(x_P) = 0$ at $x = a$ and $x = -a$. The function is decreasing on the interval $(-\infty, 0)$ and increasing on the interval $(0, \infty)$. In **Q6**, moving point P will move point A_P along the locus. When you change the value of a, however, the locus itself will move so that the locations of the zeros remain at $x = a$ and $x = -a$. In **Q7**, consider the shaded region that represents the integral. Using the equation for the area of a trapezoid, the function is $A(x_P) = 0.5(2x_P + 2a)(x_P - a)$. In expanded form, $A(x_P) = (x_P)^2 - a^2$. *Challenge:* Show that the formula for the area of a trapezoid is still valid if the shaded region is composed of two triangles.

In **Q8**, you can find the distance between the two plots for different values of a by subtracting the function A with two different values of a. Consider the first example, where $f(x) = 3$. Subtract $A(x_P) = 3(x_P - a_1)$ from $A(x_P) = 3(x_P - a_2)$. The distance between these functions is $|3a_1 - 3a_2|$. When rewritten in the form $3(a_1 - a_2)$, you can see that this difference is equal to the area of the rectangle between a_1 and a_2. In general, the difference between the plots is always equal to the area under the plot between the two values of a, because for any function, $\int_a^b f - \int_c^b f = \int_a^c f$. In **Q9**, the function $A(x)$ will increase where $f > 0$, and decrease where $f < 0$. The locations where the function will have a zero depend on the value of a, but occur at regular intervals approximately equal to π. In **Q10** through **Q13**, the function will oscillate similarly for values of $x < a$. The family of curves for different values of a are a family of trigonometric functions. The function $f(x) = \cos(x)$ matches the shape of the function for all values of a (ignoring the vertical translation.) The function A must have a zero at the point $x = a$, so the expression $-\cos(x) - (-\cos(a))$ will match the iterated plot.

In **Q14**, where $f > 0$, $A(x)$ is increasing, and where $f < 0$, it is decreasing. Where f changes sign, $A(x)$ has a maximum or a minimum. Where f is increasing, $A(x)$ is concave up, and where f is decreasing, it is concave down. Where f has a maximum or a minimum, $A(x)$ has a point of inflection. In **Q15**, where $A(x)$ is increasing, f is positive, and where $A(x)$ is negative, it is decreasing. Where $A(x)$ has a maximum or minimum, f changes sign. Where $A(x)$ is translated vertically f is unchanged.

Explore More

In **Q1**, the segment you construct has a positive slope where the signed area of the constructed trapezoid is positive, and a negative slope where it is negative. In **Q2**, where the function f is decreasing, the signed area of the next trapezoid is less than the previous, causing the slope of the segment to be less. In **Q3**, the plot has a maximum or minimum where the sign of one trapezoid is different from that of the next.

Extension

Some AP questions ask for a sketch of $\int_a^x f$ given a graph of f. Find such a question and apply the ideas in the Explore More section to help you respond to the question. See the document **Areatools.gsp** for a tool you can use to plot the integral when your function is not defined by an algebraic expression.

Getting Down to Fundamentals (page 136)

Prerequisites: Familiarity with integrals, derivatives, and antiderivatives in graphical, symbolic, and verbal perspectives. Completion of the last two activities is highly recommended.

Sketchpad Proficiency: Intermediate. Tools used in this activity were introduced in earlier activities.

Activity Time: 50 minutes

Document: Fundamentals.gsp

General Notes: This activity investigates the Fundamental Theorem from multiple perspectives.

Sketch and Investigate

In "The Area Function," steps 14–23 detailed the construction of an approximate plot of

$A(x) = \int_a^x f$ using rectangles. The difference in the y-coordinates of the constructed points P and L is equal to $h \cdot f(x_P)$, and the difference in the x-coordinates is h. You then traced a plot of the integral function using points P and L. The slope of segment PL, $h \cdot f(x_P)/h$, is equal to $f(x_P)$. So between any two points on the plot of this kind of approximation of $A(x)$, the slope of the secant is equal to $f(x_P)$. So for this approximation of $A(x)$ using rectangles, the "derivative" of A must be f, at the discrete points upon which the approximation is built.

In the previous activity, you used trapezoids, and this approximation is used here in steps 11–14. When you plot the slope of this approximation, you get a series of steps that approximate f.

In **Q1** and **Q2**, as the steps become smaller in size, they all approach the value of $f(x_A)$. The approximation of $A(x)$ was built by adding the area of a trapezoid with height h and bases $f(x_A)$ and $f(x_A + h)$ to the y-coordinate of point A.

In this construction, the y-coordinate of point B is $y_A + (1/2)(f(x_A) + f(x_A + h))h$. When you use the **Secant Slope** tool, you plot the slope of segment AB. The slope of this segment is determined by $y_B - y_A$, which equals $y_A + (1/2)(f(x_A) + f(x_A + h))h - y_A$. Because $x_B - x_A = h$, the slope of segment AB is just $\dfrac{0.5(f(x_A) + f(x_A + h))h}{h}$, which for $h \neq 0$ is the average of $f(x_A)$ and $f(x_A + h)$. As $h \to 0$, this average approaches $f(x_A)$.

Both of these approximations of $A(x)$ suggest that $A'(x) = f(x)$.

In **Q3** through **Q5**, you can show that this is true for the actual function $A(x)$. In **Q3**, because integrals on adjacent intervals "add", they also subtract. $A(x_P + h)$ is the integral from $x = a$ to $x = x_P$, and $A(x_P)$ is the integral from $x = a$ to $x = x_P + h$. The areas defined by those integrals overlap on an area equal to the integral from $x = x_P$ to $x = x_P + h$. In **Q5**, because this area, and also $A(x_P + h) - A(x_P)$, can be approximated using the formula for the area of a rectangle, the expression $f(x_P)h$ can replace the number of $\lim_{h \to 0}(A(x_P + h) - A(x_P)/h)$. The expression becomes $\lim_{h \to 0}(f(x_P)h/h) = f(x_P)$.

Explore More

In **Q1**, the points where the horizontal segments intersect the plot of the derivative are the points where the Mean Value Theorem is true. At these points, the average rate of change of f is equal to the instantaneous rate of change. For an interval $[x_0, x_1]$ this can be written $\dfrac{f(x_1) - f(x_0)}{x_1 - x_0} = f'(c_1)$. In **Q2** and **Q3**, when you move your sample points to these locations, the area of the rectangle on the interval $[x_0, x_1]$ is $f'(c_1)(x_1 - x_0)$, which is equal to $f(x_1) - f(x_0)$, the change in f on the interval. If you construct rectangles in this way, starting at a point $x = a$ and ending at $x = b$, the total area of the rectangles will be the total change in f on $[a, b]$. To show that the integral is equal to $f(b) - f(a)$, you can show that a Riemann sum formed the way you made it here falls between an upper and lower sum and so has the same limit as the width of the intervals approaches 0. In **Q4**, your explorations show that you can find the exact value of an integral of f by finding the net change in the antiderivative of f. In **Q5**, g is an antiderivative of f. Its derivative is f, so $g'(x) = f(x)$ for all x.

Extension

Make your own document that demonstrates that the function $g = \int_a^x f'$ does not necessarily equal the derivative of the function $A(x) = \int_a^x f$.